REPORTS AND STUDIES No. 61

IMO/FAO/UNESCO-IOC/WMO/WHO/IAEA/UN/UNEP
Joint Group of Experts on the
Scientific Aspects of Marine Environmental Protection
- GESAMP -

THE CONTRIBUTIONS OF SCIENCE
TO INTEGRATED COASTAL MANAGEMENT

FOOD AND AGRICULTURE ORGANIZATION OF THE UNITED NATIONS

Rome, 1996

NOTES

1. GESAMP is an advisory body consisting of specialized experts nominated by the Sponsoring Agencies (IMO, FAO, UNESCO-IOC, WMO, WHO, IAEA, UN, UNEP). Its principal task is to provide scientific advice concerning the prevention, reduction and control of the degradation of the marine environment to the Sponsoring Agencies.

2. This study is available in English only from any of the Sponsoring Agencies.

3. The report contains views expressed by members of GESAMP who act in their individual capacities; their views may not necessarily correspond with those of the Sponsoring Agencies.

4. Permission may be granted by any one of the Sponsoring Agencies for the report to be wholly or partly reproduced in publications by any individual who is not a staff member of a Sponsoring Agency of GESAMP, or by any organization that is not a sponsor of GESAMP, provided that the source of the extract and the condition mentioned in 3 above are indicated.

Cover photo: Coastline in the Daintree Area of Far North Queensland, Australia.
Courtesy of Great Barrier Reef Marine Park Authority, Townsville, Australia

ISBN 92-5-103856-2

© UN, UNEP, FAO, UNESCO, WHO, WMO, IMO, IAEA 1996

For bibliographic purposes, this document should be cited as:

GESAMP (IMO/FAO/UNESCO-IOC/WMO/WHO/IAEA/UN/UNEP Joint Group of Experts on the Scientific Aspects of Marine Environmental Protection). 1996. The contributions of science to coastal zone management. Rep.Stud.GESAMP, (61):66 p.

PREPARATION OF THIS STUDY

This study has been prepared on the basis of the work of the GESAMP Task Force on Integrated Coastal Management, established by the 24th Session of GESAMP, New York, 21-25 March 1994.

A formal meeting of the Task Team was held in Rome, 28 November to 2 December 1994, which reported to the 25th Session of GESAMP, Rome, 24-28 April 1995. It reviewed experience with the application of integrated approaches to coastal management, based on a number of case studies prepared by its members. Building upon that work, four additional case studies that focused only on mature programmes, were commissioned and later reviewed at two meetings of the chairmen of the Task Force with selected experts, held in Oslo, 11-15 December 1995 and in Rome, 12-16 February 1996. This report was completed at these meetings and subsequently reviewed by the 26th session of GESAMP, Paris, 25-29 March 1996, and approved for publication in its present form.

Contributions to the work of the Task Force by the following experts are acknowledged with appreciation: Richard G.V. Boelens, Robert E. Bowen, Chua Thia-Eng, Ingwer J. De Boer, Danny L. Elder, Edgardo Gomez, John S. Gray (Co-chairman), Graeme Kelleher, William Matuszeski, Liana McManus, Heiner Naeve (Secretariat), Magnus Ngoile, Stephen B. Olsen (Co-chairman), Jayampathi I. Samarakoon, Randell G. Waite and Helen T. Yap.

The work of the Task Force was jointly sponsored by the United Nations (UN), the United Nations Environment Programme (UNEP), the Food and Agriculture Organization of the United Nations (FAO), the United Nations Educational, Scientific and Cultural Organization - Intergovernmental Oceanographic Commission (UNESCO-IOC), the World Meteorological Organization (WMO), the International Maritime Organization (IMO) and the World Conservation Union (IUCN). The Secretariat was provided by FAO.

The Terms of Reference for the Task Team on Integrated Coastal Management were as follows:

(1)   Present a concise description of the structure of ICM emphasizing its scope and objectives;

(2)   Identify and evaluate the scientific elements (social and natural) required to support the stages of the ICM process drawing on an analysis of ICM case studies;

(3)   Identify factors and approaches that have either facilitated or impeded the incorporation of science into ICM.

EXECUTIVE SUMMARY

In this report, GESAMP draws on experience from programmes in different geographic and socioeconomic settings to identify how science and scientists can contribute to the effectiveness of Integrated Coastal Management (ICM).

The goal of ICM is to improve the quality of life of human communities who depend on coastal resources while maintaining the biological diversity and productivity of coastal ecosystems. Thus, the ICM process must integrate government with the community, science with management, and sectoral with public interests in preparing and implementing actions that combine investment in development with the conservation of environmental qualities and functions.

In the opinion of GESAMP, successful ICM programmes will involve:

a) public participation whereby the values, concerns and aspirations of the communities affected are discussed and future directions are negotiated;

b) steps by which relevant policies, legislation and institutional arrangements (i.e., governance) can be developed and implemented to meet local needs and circumstances while recognizing national priorities;

c) collaboration between managers and scientists at all stages of the formulation of management policy and programmes, and in the design, conduct, interpretation and application of research and monitoring.

From its consideration of existing experience on ICM structures and procedures, GESAMP has derived a conceptual framework to identify for each stage in the management process, the necessary contributions from natural and social scientists. GESAMP recognizes that progress towards sustainable forms of coastal development will be achieved by ICM programmes that cycle repeatedly through the stages of the management process. Each cycle may be considered a generation of an ICM programme.

It is clear that the management of complex ecosystems subject to significant human pressures cannot occur in the absence of science. The natural sciences are vital to understanding ecosystem function and the social sciences are essential to elucidating the origin of human-induced problems and in finding appropriate solutions. The need to design studies in accordance with clearly-stated objectives is particularly important. Scientific techniques and procedures that are particularly useful to ICM include resource surveys, hazard and risk assessments, modelling, economic evaluations and analyses of legal and institutional arrangements. Scientific support is also needed in the selection of management control measures and in preparing material for public information and education.

Despite great differences in the social, economic and ecological conditions in the countries from which the four case studies were drawn, there is remarkable consistency in the lessons learned about the contributions of science to ICM. They demonstrate that scientists and managers must work together as a team if scientific information generated for ICM is to be relevant and properly applied for management purposes. Since the two professions have different perspectives and imperatives and approach the solution of problems differently, the objectives and priorities for programmes must be derived, tested and periodically re-evaluated by scientists and managers working together.

GESAMP recognizes the need to build constituencies for ICM initiatives and the importance of matching policies and management actions to the capabilities of the institutions involved. Some countries experiencing severe coastal degradation and where remedial measures are urgently required, may not have the necessary frameworks for environmental management and must focus much of their effort initially on creating the institutional context in which effective resource management can occur.

CONTENTS

ABSTRACT

GESAMP (IMO/FAO/UNESCO-IOC/WMO/WHO/IAEA/UN/UNEP Joint Group of Experts on the Scientific Aspects of Marine Environmental Protection)
The contributions of science to coastal zone management.
Reports and Studies, GESAMP. No. 61. Rome, FAO. 1996. 66 p.

The scope, objectives and defining features of Integrated Coastal Management (ICM) are briefly described and a conceptual framework for the effective operation and evolution of ICM programmes is presented. ICM is a dynamic and continuous process by which progress towards sustainable use and development of coastal areas may be achieved. ICM programmes therefore have the dual goals of conserving the productivity and biodiversity of coastal ecosystems while improving and sustaining the quality of life of human communities. This requires the active and ongoing involvement of the interested public and the many sectoral groups with interests in how resources are allocated, development options are negotiated and conflicts mediated.

Selected case studies from a diversity of settings in developed and developing nations reveal striking commonalties in the interplay between science and ICM and demonstrate that effective ICM cannot occur in the absence of science. The natural sciences are vital to understanding the functioning of ecosystems and the social sciences are essential to comprehending patterns of human behaviour that cause ecological damage and to finding effective solutions. Scientists and resource managers often have different perspectives and imperatives. Nevertheless, as the case studies clearly suggest, they must work together as a team through all stages of an ICM programme and reach agreement on the scientific work needed to address priorities and guide policy development. The case studies also underscore that programmes must tailor their scope and objectives for a given period to the capabilities of the institutions involved. Where such institutions are weak, and the constituencies required to support an ICM initiative are not yet in place, programmes must first work to create societal conditions wherein the community will be receptive to the aims and procedures of resource management.

Key Words: coastal management, science and public policy, policy process, public participation, resource management.

1. INTRODUCTION

In a number of previous reports, GESAMP discussed the priorities for global action in managing coastal and marine environments and the contributions that science can make to these important tasks. The following extracts from the report of the Twentieth Session of GESAMP (GESAMP, 1990) are particularly relevant to the present document:

> *"The concept of sustainable development implies that the present use of the marine environment and its resources shall not prejudice the use and enjoyment of that environment and its resources by future generations. Past practices that have neglected this principle are the fundamental cause of many current environmental problems."*

> *"Development inevitably implies environmental change. The challenge for marine and coastal zone management is to balance short-term development needs against long-term sustainability of ecosystems, habitats and resources such that the range of choices and opportunities available to future generations is not diminished by the consequences of present development choices."*

> *"Comprehensive area-specific marine management and planning is essential for maintaining the long-term ecological integrity and productivity and economic benefit of coastal regions."*

> *"The effectiveness of management actions to protect the ocean cannot be assessed without scientific analysis and knowledge. Accordingly, comprehensive protection strategies should incorporate scientific principles; however, it is recognized that decision-making frequently involves considerations other than scientific arguments. Close interaction among scientists and decision-makers is essential."*

At the international level, much attention has been given to articulating the need for Integrated Coastal Management (ICM), the scope of ICM programmes and the issues they should address. Relevant documents from international fora include Chapter 17 of Agenda 21 of the United Nations Conference on Environment and Development (United Nations, 1993), the Noordwijk Guidelines for Integrated Coastal Zone Management (World Bank, 1993), the report of the World Coast Conference (IPCC, 1994) and numerous technical reports released by international organizations, including UNEP (1995), FAO (Clark, 1992; Boelaert-Suominen and Cullinan, 1994), OECD (1993) and IUCN (Pernetta and Elder, 1993). Several GESAMP reports (e.g., GESAMP, 1980; GESAMP, 1991a; GESAMP, 1994) have addressed the interrelationships between the condition of coastal and marine environments and human activities.

Coastal and marine environments are particularly vulnerable to over-exploitation because they include large areas traditionally considered to be "commons". Before and since Garrett Hardin's essay *The Tragedy of the Commons* (Hardin, 1968), there has been ample evidence that the long-term effect of uncontrolled human activity on the commons is usually to degrade or destroy it. Furthermore, coasts often include areas where a diversity of incompatible activities compete for limited space and resources. The profits and benefits of some activities are confined to minorities, while costs are imposed on the community and the environment.

Although a clear understanding of the factors involved is often lacking, widespread concern over the condition of coastal environments has led to demands by the public for the right to participate in decisions affecting the coast and for better protection of coastal resources. As a result, there has been parallel development of ICM programmes in various parts of the world that actively involves the public in improving the management of coastal areas. In economic terms, these methods aim to ensure that the costs generated by one sector of society are not imposed on the community generally. The four case studies examined in this report attest to the centrality of this public process.

This report is offered as guidance to those responsible for the oversight and funding of ICM programmes, those engaged in the design and implementation of programmes and the natural and social scientists who participate in the ICM process. While the experience upon which this report is based represents a wide range of settings and approaches it underscores the many commonalities in the factors that influence how the sciences can contribute to ICM programmes and thus affect the success of these initiatives.

2.   THE OBJECTIVES AND SCOPE OF INTEGRATED COASTAL MANAGEMENT

2.1   Objectives

Integrated Coastal Management (ICM) is a process that unites government and the community, science and management, sectoral and public interests in preparing and implementing an integrated plan for the protection and development of coastal ecosystems and resources. **The overall goal of ICM is to improve the quality of life of human communities who depend on coastal resources while maintaining the biological diversity and productivity of coastal ecosystems.**

Expressed in this way, the goal of ICM is clearly consistent with national and international commitments to **sustainable development** for all environments (terrestrial and marine), from the headwaters of catchments (watersheds) to the outer limits of exclusive economic zones (EEZ), whether or not they are subject to multiple jurisdiction.

Central to success in achieving this goal is the need for ICM to provide an equitable, transparent and dynamic governance process that is acceptable to the community.

2.2   ICM in the Context of Environmental Protection and Management

In a previous report on *Global Strategies for Marine Environmental Protection* (GESAMP, 1991a), GESAMP presented a framework for environmental protection and management that provides for the various political, social and scientific inputs that are needed in developing programmes to protect the environment and to ensure the sustainable use of natural resources. The framework is applicable to all sectors of the environment, terrestrial, freshwater and marine. Environmental management is, therefore, an implicitly **holistic** process and the approach to managing coastal areas is fundamentally the same as that which should be used to manage a nation's environmental heritage in its entirety.

From its analyses of environmental problems confronting coastal areas and communities of the world, including those highlighted by the case studies reviewed in this report *(see Annexes 1-4)*, we conclude that a majority of ICM programmes will need to deal with one or more of the following three conditions:

•   **Over-exploitation** of renewable resources, either directly by harvesting or by the destruction or modification of habitats and disruption of predator/prey and other ecological relationships;

•   **Conflicts** that arise where several human activities that depend on the same area and/or resource are incompatible;

•   **Insidious damage**, including loss of biological productivity and diversity, that may result from cumulative impacts of different practices.

Table I
The Scope and Focus of ICM Programmes

Chapter 17.5 of Agenda 21 describes the scope and process of ICM programmes. The text calls for programmes that:

- identify existing and projected uses of coastal areas with a focus upon their inter-actions and interdependencies;

- concentrate on well-defined issues;

- apply preventive and precautionary approaches in project planning and implement-ation, including prior assessment and systematic observation of the impacts of major projects;

- promote the development and application of methods such as natural resource and environmental accounting that reflect changes in value resulting from uses of coastal and marine areas;

- provide access for concerned individuals, groups and organizations to relevant information and opportunities for consultation and participation in planning and decision-making.

2.3 Principal Features of ICM

ICM is a continuous and dynamic process that addresses the use, sustainable development and protection of coastal areas. ICM requires the active and sustained involvement of the interested public and the many stakeholders with interests in how coastal resources are allocated and conflicts are mediated. The ICM process provides a means by which concerns at local, regional and national levels are discussed and future directions are negotiated. The concept of an **integrated** approach to the management of coastal areas is intentionally broad and has four elements:

Geographical: It takes account of interrelationships and interdependencies (viz., physical, chemical, biological, ecological) between the terrestrial, estuarine, littoral and offshore components of coastal regions;

Temporal: It supports the planning and implementation of management actions in the context of a long-term strategy;

Sectoral: It takes account of interrelationships among the various human uses of coastal areas and resources as well as associated socio-economic interests and values;

Political/Institutional: It provides for the widest possible consultation between government, social and economic sectors and the community in policy development, planning, conflict resolution and regulation pertaining to all matters affecting the use and protection of coastal areas, resources and amenities.

The emphasis on integrated management means that ICM programmes should:

- encourage an **interdisciplinary analysis** of the major social, institutional and environmental issues and options affecting a selected coastal area followed by a decision on the issues that should be addressed within a given period. The analysis should take into account the interactions and interdependencies among natural resources and different economic sectors. An ICM process must consider all relevant practices in a given locale—typically including fisheries, aquaculture, agriculture, forestry, manufacturing industry, waste disposal and tourism—in the context of the needs and aspirations of the communities affected. It should distinguish between issues that are likely to be important over long time-scales (e.g., climate change, population growth and the consumption habits of society) and more immediate concerns such as those associated with the governance process, conflicts among user groups and current social, economic and environmental conditions.

- initiate a **dynamic policy process** that is explicitly designed to evolve through experience, rather than an inflexible plan that provides for a limited set of responses to immediate problems. This requires continuous improvement of the information base, ongoing assessment of policies, administrative arrangements and options for problem resolution, and a robust administrative system. Such learning and adaptation requires the sustained monitoring and evaluation of trends in the condition and use of the ecosystems in question as well as the effectiveness of governance responses in order to periodically refine the design and operation of the programme.

- provide a **formalized governance structure and set of procedures** to provide continuity and to maintain confidence in the management process. ICM programmes are most likely to build and maintain active constituencies within the societies affected when the planning and decision making process is transparent and participatory. The programme must be accountable for its actions and must demonstrate that it has the capacity to resolve conflicts and implement its policies and plans. Without strong constituencies both within central government and at the local level, no ICM programme can be both effective and sustainable.

- promote **concern for the equity issues** posed by existing methods of resource allocation. The maintenance of critical stocks of natural resources, ecosystem processes and environmental qualities are goals that transcend the present and require consideration of the benefits and opportunities that should be available to future generations.

- commit to **making progress towards the goal of sustainable development** and therefore achieving a balance between both development and conservation. ICM must aim to combine and harmonize investment in development with conservation of environmental qualities and functions. This is because human populations share a common suite of needs and demands that include employment, housing, education, health care and basic utilities as well as a healthy natural resource base that can maintain the goods and services that sustain communities. In most cases an ICM programme cannot define or achieve sustainable levels of development in a single step. Progress will be made only by maintaining a programme through a series of generations, each of which is marked by the completion of the five stages in the ICM process *(see Section 3)*.

Once formally adopted, ICM programmes have institutional identity typically granted by legislation or an executive mandate. Formalized ICM programmes therefore have continuity as independent organizations or as a programme administered through a network of organizations. In both cases' roles and responsibilities for planning and implementation are clearly delineated. The institutional structure typically contains distinct but clearly linked mechanisms for (i) achieving interagency coordination at the national or regional level (e.g., through an interministerial commission, authority or executive council) and (ii) providing for conflict reduction, planning and decision-making at the local level.

2.4 Boundaries and Scale

Ideally, the geographic boundaries for an ICM initiative should encompass a stretch of coast and adjacent ecosystems that are linked by common natural (e.g., climatic, physical, biological) features and/or by the occurrence of particular human activities. This would include those terrestrial systems that significantly affect the sea, or are affected by their proximity to the sea, and those marine systems affected by their proximity to the land; it implies boundaries that (a) include those areas and activities within watersheds that significantly affect the coast, and (b) may, in certain cases, extend seaward to the edge of the continental shelf or the Exclusive Economic Zone (EEZ).

In practice, the boundaries of first generation ICM programmes *(see below)* are often determined by the specific issues that the programme selects for its initial focus. For example, a programme that is initially most concerned with issues of coastal erosion and tourism development might reasonably adopt boundaries that are narrower than those of a programme concerned with water quality or fisheries.

Related to the problem of boundaries is the question of scale. ICM programmes usually cover geographic areas within a particular country rather than the whole country or only parts of whole ecosystems as in the case of a bay or watershed shared by two or more countries. The area addressed by an ICM programme may be large or small but the boundaries set should suffice for most local management decisions. Decisions and actions required in addressing the needs of the region may transcend the delineated boundaries. Furthermore, decisions made at a higher political or national level often have great significance for the area being managed.

The question of scale is particularly important for communities that rely on resource exploitation in a particular area. Once the requirements of the population exceed the ecosystem productivity, the manager must consider external subsidies or the need for alternative resources if the consumption rate or quality of life of the community is to be maintained. The alternative is to reduce demand for the resources. Reducing the population by emigration is seldom practical.

3. THE CONTRIBUTIONS OF SCIENCE TO THE STAGES OF AN ICM PROGRAMME

The papers referenced in Section 1 provide detailed descriptions and diagrams of the steps in the ICM process and others (e.g., Chua and Scura, 1992) have provided conceptual frameworks for linking management processes and options with specific issues. The simplified sequence of stages presented here is consistent with, and draws from, these publications and focuses on the contributions of science. The stages are summarized in Figure 1 and discussed in greater detail below. As the figure clearly illustrates, these five consecutive stages form an ongoing, iterative process that may go through a number of cycles before the programme is sufficiently refined to produce effective results. Each completion of the five stages may be termed a generation of a programme.

The types of scientific support required by ICM evolve with each stage in the process; a synopsis of the main scientific inputs at each stage follows. Additional information on selected natural and social science techniques and approaches is given in Section 4.

3.1 Stage 1: Issue Identification and Assessment

This is where the requirements of an ICM programme are initially defined and assessed. It is essentially a process of compiling, integrating and prioritizing information that defines the environmental, social and institutional context within which the ICM programme will proceed. The major topics to address are as follows:

Assessment of the condition of coastal systems:

• characterization of significant habitats, species and biological communities, living and non-living resources and their interrelationships;

- identification of trends in the condition and use of resources and amenities;

- estimation of short and long-term implications of such changes for the environment and society;

- identification of particular sub-areas and conditions that warrant priority within the ICM programme.

Assessment of the policy and institutional context:

- roles and responsibilities of agencies as they relate to priority ICM issues;

- assessment of institutional capability, capacity and credibility for addressing these issues;

- identification of existing policies and goals relevant to these issues.

Assessment of the development context:

- assessment of trends in quality of life indicators;

- identification of stakeholders for priority ICM issues, their values and interests;

- initial assessment of societal perceptions of the issues and their implications.

Clearly, Stage 1 is crucial because it provides the foundation for subsequent stages in the process that leads to a full-fledged ICM programme. Despite the range of information to be compiled and assessed, it should be possible to carry out Stage 1 within a period of 6-18 months.

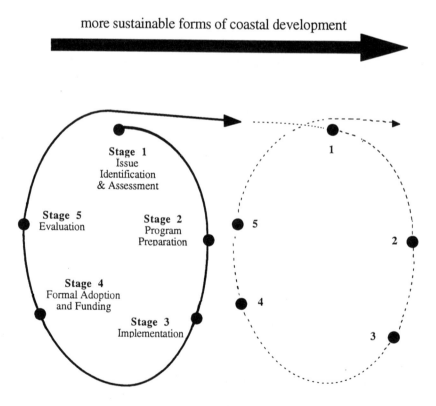

Figure 1. The stages of the ICM cycle to which sciences contribute.
The dynamic nature of ICM requires feedbacks among the stages and may alter the sequence, or require repetition of some stages.

Scientific input to Stage 1

The process of sorting through and assessing large amounts of information of variable quality on a wide range of topics requires skill and judgment. The assistance of natural and social scientists, preferably those familiar with local and national circumstances, will be needed to find existing information, to assess its relevance and quality and to clearly define and prioritize the issues to be addressed. The linkages between issues must be defined and evaluated. Stage 1 should also identify any obvious gaps in scientific knowledge, their likely implications for the ICM process and the practical possibilities for filling them within a realistic time-frame.

A team of natural and social scientists should participate in the participatory process of preparing a document (e.g., a programme profile) that describes, in general terms, the various issues on which the programme will focus and the associated values, policies or constraints under which the programme will operate. Such scoping documents should identify the long and short-term implications of existing trends and suggest priorities for action. Such documents are an essential basis for consultations among managers, scientists and the public at large on the goals and priorities of the programme.

3.2   Stage 2: Programme Preparation

In contrast to the relatively rapid assessments of Stage 1, this stage involves a more protracted consultative and planning process that evaluates different options for action. This process may take several years. The main purpose is to develop a management plan that constitutes "a vision for the future" and that expresses, in realistic and tangible terms, the qualities of the environment to be achieved and maintained, the way in which resources should be allocated and any necessary changes in patterns of resource use and human behaviour. During this stage the specific objectives of the programme must be clearly defined. These should reflect the aspirations and values of those with an interest in the areas and resources to be managed. It is important to ensure that this process of planning and evaluation of options provides sufficient time for meaningful incorporation of stakeholders at the community level such that constituencies are built that will actively support the management objectives and strategies that are selected as the programme's focus.

Since the planning process is complex, continuing for several years and involving large numbers of people, the best approach may be to generate and test a variety of strategies and objectives, thereby building confidence in, and support for, a decreasing number of options. Thus, the planning process may involve several iterations comprising analysis, debate and pilot-scale implementation as the ICM team explores the feasibility of alternative management actions and associated governance requirements.

During the initial (viz., first generation) planning cycle, it may be advisable to focus on a few, relatively small-scale, areas where management policies and techniques can be implemented and to postpone attempts to manage the entire coastline until subsequent generations of the programme. This is often the most responsible approach to dealing with a crisis, such as coral reef blasting or mangrove destruction, where some early and visible action may be needed pending research to find the optimum solution.

Scientific input to Stage 2

Natural and social scientists should be well represented on the ICM team during the programme planning phase both to explain and expand on the findings of the Stage 1 assessments and to assist in defining and planning studies to fill important gaps in information. It is especially important that research be initiated early in the programme to address:

• the characteristics and conditions of coastal systems that cause concern or otherwise warrant attention;

Table II
Scientific questions relevant to the destruction and restoration of coastal habitats

---

What is the scale of habitat destruction?

This is logically the first question to be addressed. Subsequent management and scientific action would be dictated by perception of the magnitude of the problem. Detection of the scale of habitat destruction is aided by modern technological tools, such as remote sensing, acoustic surveys of sediments and Geographic Information Systems (GIS). Reference to historical records is often indispensable, as is anecdotal evidence.

What are the natural processes that maintain habitat integrity?

Intelligent resource use, including land use, and planning and zonation, is premised on a knowledge of natural processes that could lead to alterations in habitat characteristics such as topography and productivity over the long-term.

What are the dynamic linkages among habitats that need to be considered in maintaining sustainable use of their resources?

Habitats which are spatially separated are often dependent on each other for the exchange of material and energy. An example is the recruitment of important species of reef fish and probably coral larvae from other habitats or areas which serve as nursery areas.

Can the links between habitat degradation and human activities be quantified?

This involves studies on the characteristics of human activities that relate to alterations in the national resource base (growth, migration, decline), such as changes in patterns of resource use and in the application of technology to exploit resources.

How many species are actively dependent on the habitats concerned?

Are all species equally important for conservation purposes?

What are the spatial and temporal scales of natural habitat recovery?

When management appreciates the areal extent and length of time involved for degraded habitats to recover naturally, decisions must be made regarding the necessity of intervention. Insights into the natural processes underlying recovery (including spatial and temporal scales) may be derived from natural events, e.g., observations on recolonization following a large scale disturbance (e.g., a hurricane).

Which species play a key role in natural recovery process?

Particular species play more critical roles than others in terms of maintaining ecosystem functions (e.g., productivity, nutrient cycling, predation). Such knowledge will guide restoration strategies which, for logistic reasons, may have to focus on the minimum complement of species to achieve natural recovery.

- the governance process itself, i.e., the decision-making process, compliance with voluntary or regulatory controls on various practices and/or behaviour;

- the factors and processes that regulate these characteristics and conditions.

Scientists should work with managers to prepare concise statements of objective for research and monitoring, clearly defining what is to be measured and why, and in identifying methodologies, facilities and personnel needed for the studies to be cost-effective and successful. For each priority issue to be addressed, scientists and managers should together formulate specific questions that are to be resolved through subsequent scientific investigations. Questions relevant to one particularly common issue, the destruction and restoration of coastal habitats, are given in Table II.

Specific Stage 2 tasks to be addressed by the ICM team and to which natural and social scientists should contribute, typically include:

- estimating the relative influences of anthropogenic and natural factors in causing particular changes to coastal systems and resources;

- characterizing the likely short and long-term consequences for society of existing trends in the condition and use of the coastal environment;

- assessing the social and economic benefits which stem from various options for the use and exploitation of coastal resources and amenities;

- formulating approaches to mitigate or reverse environmental degradation;

- estimating the social, environmental and economic costs and benefits of alternative actions.

3.3   Stage 3: Formal Adoption and Funding

Formal adoption of a programme will generally require a high-level administrative decision, for example by the head of a government agency, a minister or the cabinet, or perhaps by presidential endorsement. It will include consideration and agreement of a budget (i.e., levels and sources of funding) for each phase of the programme. A phased budget has certain advantages. For example, a preliminary budget could be adopted to allow the scientific research and baseline monitoring developed in Stage 2 to be initiated in advance of other programme elements.

It is to be expected that plans for ICM programmes will be subject to detailed scrutiny and questioning, and will often need revision, before they are formally approved. Consequently, this stage may be characterized by a dramatic change from the technical to political aspects of the ICM process. The interests of governmental agencies and commercial sectors affected by the programme may give rise to new and unexpected arguments that the ICM team must address. Similarly, formal approval often does not guarantee adequate funding. Securing the funds required for implementation of an ICM plan may require another round of planning to review possibilities for cost reduction and increased efficiency and perhaps a slower rate of implementation. The process is one of bargaining and accommodation.

Scientific input to Stage 3

Access to scientific advice is useful, and sometimes essential, when attempting to react quickly to the issues that emerge during the political bargaining process. Topics that are typically closely examined and challenged, and that involve natural and social sciences are:

- cost/benefit and decision analysis;

- arguments over whether the proposed actions can be reasonably expected to produce the results being promised, both in changed behaviour and the condition of the ecosystem.

3.4   Stage 4: Implementation

At this stage in the ICM process, the management plan becomes operational and the emphasis shifts to the introduction of new forms of resource development and use, new institutional arrangements and monitoring systems and the application of new controls, regulations and incentives.

Enforcement is an essential element of programme implementation and one which clearly demands a constant supply of reliable and readily interpretable monitoring data.

Successful implementation of an ICM programme invariably presents new, sometimes unforeseen, challenges and the ICM team must be able to respond to these while maintaining momentum within the core programme. Priority activities during this stage typically include:

- conflict resolution;
- inter-agency coordination;
- infrastructure construction;
- development actions;
- public education;
- training of management or enforcement personnel;
- planning and research on new areas or problems.

Scientific input to Stage 4

From this stage, monitoring should be re-focused to measure changes in the areas and resources to be protected, the practices to be modified and changes in the forms of development that the programme seeks to provide. Such monitoring must be designed to generate data that can be compared with results from the baseline studies and to the specific development and conservation objectives contained in the programme plan. The design, implementation and management of these studies is an essential function for the ICM team and its supporting scientific bodies and advisors.

Key roles for natural and social scientists at this stage are to assist other ICM team members in translating information from monitoring programmes and assessing the efficacy of new measures. This is part of the learning process and is particularly important where management techniques and approaches are tested on a pilot scale. Scientists should test the hypotheses developed in Stages 1 to 3 and on which the programmes' actions are based. They should also give advice on whether elements of the programme should be revised or adapted to improve their effectiveness or efficiency and on developing new technologies that help attain programme objectives. This process of learning and adaption should continue throughout the implementation stage and is nourished by scientific knowledge and skills.

3.5   Stage 5: Evaluation

This stage, where the greatest learning should occur, has been omitted or performed in a superficial manner in a great majority of coastal management initiatives. Yet, if ICM programmes are to proceed through a series of generations to more sustainable forms of coastal development, this stage should be the critical juncture between one generation and another. The evaluation stage must address two broad questions:

What has the preceding generation of the programme accomplished and learned and how should this experience affect the design and focus of the next generation?

How has the context (priority issues, environment governance) changed since the programme was initiated?

This, in essence, sets the stage for repeating the assessments in Stage 1.

A meaningful evaluation can be conducted only if the programme objectives have been stated in unambiguous terms and if indicators for assessing progress were identified in Stages 2 and 3, and monitored during the preceding generation. Baseline data are essential. Many evaluations yield ambiguous results because these preconditions for assessing performance do not exist.

Scientific input to Stage 5

Natural and social scientists have important roles to play in the programme evaluation process. In particular, they should evaluate the relevance, reliability and cost-effectiveness of scientific information generated by research and monitoring and advise on the suitability of control data. Such analyses are necessary if funding agencies are to be persuaded that the often substantial investments in scientific work are justified. Scientists should also provide estimates of the extent to which observed changes in managed environments and practices are attributable to ICM measures as opposed to other factors.

3.6   Effectively Integrating Science and Management

ICM assumes continuously increasing knowledge of how ecosystems function and respond to anthropogenic forces. Equally important is an appreciation for the values and needs of the human societies in question and the capabilities and interests of the institutions that will play roles in the management process.

It must be recognised, however, that there are often important differences in the pressures and motivations acting upon the ICM manager, as opposed to the scientist collaborating in the programme. Managers must produce successes that bring credibility to the programme. They must also respond to crises and the demands of administering a complex set of activities often in the light of public and political demands for action. The scientist, on the other hand, is primarily concerned with the generation and appropriate use of "good science". To sustain a productive relationship between scientists and managers, both parties must work to achieve:

*   common support for the goals and objectives of the programme;

- mutual understanding of the respective pressures and reward systems under which scientists and managers operate;

- long-term commitment to progress of the programme;

- continuous output of information and progress reports so that all parties are aware of decisions and events affecting the course of the programme.

Each of the Case Studies provides examples of how this relationship has worked in practice.

3.7   Structures for a Relationship

A range of structures can be helpful in achieving these results. In some cases, as illustrated by the Ecuador case study *(see Annex 3)*, informal inter-institutional working groups on specific issues of concern to the programme can be productive. Informal structures may avoid a situation whereby official delegates feel obliged to articulate and defend the policies and prerogatives of their respective institutions. Scientists with an interest in the issue participate as individuals, and the emphasis is upon problem solving, finding areas of common interest and collaborative action. Some programmes rely upon more structured advisory boards comprising a cross section of natural and social scientists, and institute representatives; typically these boards will focus on giving advice on the 'technical' aspects of a programme. In other cases, technical boards or committees, in which scientists participate, are authorized to propose priorities and funding for research and monitoring to be sponsored by the ICM programme. The latter requires consideration of any existing and relevant scientific studies that are funded by outside sources.

4.   RELEVANT TECHNIQUES IN SCIENCE AND MANAGEMENT

Experience from environmental management generally has shown that there are various techniques and approaches in the fields of natural and social science that can help to improve the efficiency and effectiveness of ICM in achieving its overall goal. There are also a number of measures that managers can apply where the need to control specific practices has been demonstrated. Some of the most common techniques, approaches and measures are described below.

4.1   Research and Monitoring

Critical considerations in planning, budgeting and organizing coastal science programmes for management purposes are:

- the questions and problems that scientists will be asked to address must be explicitly stated—scientists should help in formulating the questions;

- no study should be funded or initiated unless and until its objectives have been written and recorded in clear and unambiguous terms;

- studies should seldom be undertaken in isolation; the inter-dependencies and inter-relationships between studies should be recognized and mechanisms established for efficient co-ordination and collaboration.

4.2   Integration of Research, Monitoring and Assessment

Without careful planning and proper coordination, research and monitoring can dissipate large amounts of money while failing to provide the information that is most needed for environmental management. Relevant and cost-effective information is most likely to result from studies that are initiated as part of a planned and well integrated programme. Nevertheless, information from scientific studies of an area, despite their different origins and purposes, are often useful to ICM initiatives.

An integrated programme for coastal science will encompass basic and applied **research**, **monitoring** (i.e., repetitive measurements using validated methods) and periodic **assessments** of environmental quality. Ideally, the programme will aim to:

• provide information needed for management and protection purposes while simultaneously advancing scientific understanding of the area (i.e., its components and processes, and how these combine to form a functioning ecosystem);

• provide for continuous assessment of **environmental changes** and human contributions to these changes; and

• make the most efficient use of scientific resources, for example by maintaining high standards of project design and data quality assurance.

This implicitly suggests that the programme should adopt a medium to **long-term** work plan comprising a set of studies with **clearly-defined objectives**. The plan should take account of the **interrelationships** and inter-dependencies between the studies and the sequence in which they should be undertaken. There should be no requirement for all studies within the programme to be funded by the same agency or management initiative such as ICM.

An integrated science programme should aim to ensure that activities in research and monitoring are, as far as possible, complementary. As a general rule, all monitoring should be preceded by research to develop suitable methodologies, to clearly establish the conditions under which the methodologies can be usefully applied and to verify their suitability for routine application. An important part of this process is to ensure that sufficient knowledge exists to allow the results from monitoring to be reliably interpreted. This frequently requires a good understanding of the cause and extent of variability (both temporal and spatial) in the parameters to be monitored which, in turn, will help to determine the optimum frequencies and locations of sampling.

4.3   Objectives

Possibly the most common and serious error in environmental science is that studies are initiated before the objectives have been fully and clearly documented. This can have serious implications, particularly for monitoring programmes. Objectives dictate what should be monitored, how, when and where, i.e., they determine the **design** of the programme; they are also the main criteria for use in interpreting the results. Thus, objectives should always be drafted with great care and clarity, ideally in the form of specific questions or hypotheses, and in the knowledge that suitable methodologies are available. As a minimum, a statement of objectives for a scientific investigation should specify **what** is to be done and **why**. Proper care in formulating objectives is the key to successful and cost-efficient science.

4.4   Scope

The scope of a coastal science programme will largely be determined by the extent of existing knowledge, the data needed for compliance with existing management requirements and the priorities for new information. Thus, programmes may range from those which address a few current and important issues to those which are broad and multi-disciplinary. As recommended in Stage 1 of the ICM process (Chapter 3.1), an assessment of existing knowledge is the best way to identify important **gaps and deficiencies**. Ideally, this assessment should be an integral part of the process of preparing a report on the present condition (i.e., quality status) of the environments concerned. GESAMP has recently published guidelines on how such reports may be drafted (GESAMP, 1994).

Although the data requirements of managers and existing national monitoring programmes deserve priority, a well-balanced programme will include studies designed to progressively improve scientific understanding of how coastal ecosystems function and their responses to human intervention.

4.5   Impact Evaluation

Under this generalized heading there are a number of roles for social and natural scientists within ICM programmes. They include assessments of the **causes** and **implications** of existing environmental impacts (termed environmental auditing) and also the prediction of impacts arising from proposed development and from actions to combat existing problems. Assessments of social and economic impacts of environmental changes and new developments are increasingly necessary. Advice developed from impact evaluations is of utmost importance to managers—it is often the deciding factor in justifying the choice of action and the financial resources to commit.

For some ICM programmes, especially those funded by the development banks, an early assessment of anticipated economic and social impacts of the programme may be required. This typically involves the identification of the beneficiaries of proposed ICM actions and the particular benefits they may reasonably expect to receive.

Two well-established techniques for use in impact evaluation are **hazard** profiling and **risk** analysis. A hazard is an intrinsic and potentially harmful property; it usually applies to properties (e.g., toxicity, bioaccumulation potential, etc.) of chemical pollutants but may also be applied to human activities. For example, the uncontrolled exploitation of renewable resources or irreversible alterations to ecologically important habitats are **hazardous** because they potentially may result in long-term reductions in productivity, biodiversity and the livelihood of local people. Although it is sometimes useful to compare the hazards of different materials or activities, hazard profiles are most valuable when used in risk analysis. In environmental science, a risk is the probability (in statistical terms) that a practice involving a given hazard will result in a particular change, effect or response. Risk analysis requires information on the scale (source strength, temporal and spatial aspects) of the practice and the physical, chemical and biological characteristics of the environment in which it occurs. It also involves estimations of "downstream" impacts (side-effects); knowledge of environmental processes combined with **modelling** and/or **pathway** analysis are especially useful for this purpose.

Although impact evaluation is indispensable for resource management, it is not an exact science. Any analysis or prediction of environmental change will involve a degree of **uncertainty**. Important tasks for scientists are to minimize these uncertainties by obtaining the most accurate possible data on relevant variables and, above all, to ensure that managers are aware of the uncertainties inherent in the results or conclusions they provide.

4.6   Resource Surveys

A fundamental requirement of ICM programmes is an account of the different ecosystems, resources, geochemical features, habitats and biotic communities within the geographical boundaries of the programme. Summaries of data on social, industrial and other activities, and human uses of different environmental sectors and resources, will also be needed. Much of this information may already exist in the scientific literature and in national agencies dealing with agriculture, geology, forestry, wildlife and fisheries. It is nevertheless important that ICM scientists should, as an initial task, collate and classify this information in the form of maps, thematic directories and/or computerized databases such as **geographical information systems** that are a useful means of cataloguing and retrieving social, economic and environmental data.

How detailed and comprehensive a resource inventory should be depends on the particular focus and priorities of the ICM programme. Some first-generation programmes may focus on practices and problems that are largely confined to specific ecosystems and habitats such as estuaries, beaches or reef systems; more detailed information will therefore be required for these areas. However, as ICM programmes mature, it will be important to progressively extend resource databases and to fill critical gaps in the biology, ecology, geochemistry and other features of the larger, or more important, habitats.

The most useful resource inventories will be both **qualitative** and **quantitative**. This will allow estimation of the magnitude of changes and trends in environmental and social conditions. Clearly, for quantitative information to be reliable it must be reasonably up-to-date, and heavily exploited resources will need to be resurveyed periodically. Although many environmental features (e.g., fisheries, water quality, waste disposal) will require land or sea surveys, such features as the scale and distribution of major habitats, beach systems and shoreline developments can be more efficiently assessed by conducting **aerial photographic surveys** (normal colour and/or infra-red) or through analysis of recent **satellite imagery**.

4.7   Modelling

Numerical models can help to improve understanding of complex environmental **processes** (e.g., the movement of materials in water or the atmosphere) and to resolve questions and problems in resource management. Economic models can be valuable tools in environmental management, for example in comparing the economic consequences of different development, conservation or pollution control options. Typical applications of models in marine science include the **simulation** of seawater circulation under different seasonal, tidal and weather conditions and the **prediction** of larval dispersal, contaminant pathways and distributions. In the social sciences, economic models can be used to predict, for example, the effect of demographic changes on the demand for goods and services derived from natural resources.

In the context of ICM, the cost of sophisticated models and their reliance on high volumes of accurate data make their utility limited except in the case of advanced programmes. Modelling activities must be well supervised to ensure that they remain tightly focused and do not become the justification for extensive monitoring to generate the necessary calibration data. Models should be used only where there are well-formulated questions or hypotheses for the models to address.

The use of numerical models in studying the **transport**, **dispersion** and **fate** of contaminants in coastal regimes, and the factors to be considered for parameterization and formulation of these models, have been described in some detail in a previous report (GESAMP, 1991). A synopsis of key points in the introductory sections of this document is given below.

Models produce **estimates** whose accuracy depends on the quality of the environmental data used to **calibrate** and **validate** the model. Furthermore they depend upon how well the model simulates known processes such as turbulent dissipation of kinetic energy. No single model is appropriate for all purposes and a range of models is usually required. Successful models can frequently only be developed from a sound knowledge of the processes in the region of interest. This may require a well-conceived **observational programme** prior to the modelling work to determine scales of motions, sediment sizes and types, and other parameters. Any field programme should be designed to subsequently provide validation for the model.

Only in exceptional circumstances can complex models developed by one group be handed over to another. All models and their solutions have **limitations** that need to be understood, and this requires that a minimum of expertise be transferred with the models. Although it is difficult to produce generic models, coastal processes themselves may be considered as generic. The **parameterization** procedures that are relevant to one situation may be equally applicable to other coastal regions. It is usually the combination of processes and their relative importance that distinguishes one coastal location from another. However, parameter values themselves are often site-specific and, in general, the values of most parameters in a model have to be selected on the basis of data obtained at the site.

4.8   Economic Assessment and Valuation

An economic assessment of goods and services derived from natural systems has traditionally focused on those that have market prices, such as goods utilized as food and other marketable products (shellcraft, construction materials, etc). Such assessments attempt to incorporate **direct use value** as well as **ecological function and option values**. The values of natural systems such as the function of coral reefs in buffering wave energy that would otherwise lead to coastal erosion, the capability of mangroves to trap sediments that would otherwise remain suspended and contribute to low water quality, or the aesthetic value of natural systems that contribute to the growth of tourism, are examples of ecological functions that need evaluation. By internalizing the worth of these otherwise non-marketable functions, the economic value of environmentally derived goods and services are better estimated, and development options that protect or destroy the environment are appropriately evaluated. The economic assessment and valuation of natural systems rely heavily on what is known in terms of rates and processes relevant to their biological productivity and recovery under various stress scenarios. A major limitation of this method is that economic valuation has a relatively short time frame, typically 5 to 10 years as it considers prices, inflation, and discount rates. This time frame becomes irrelevant in terms of the growth and geomorphological evolution of resources such as coral reefs which span geologic time scales.

4.9   Legal and Institutional Analyses

Laws and institutions provide a framework by which a society organizes interactions among its people and governmental and non-governmental institutions. A thorough analysis of these is essential. As environmental laws evolve, they are not always consistent with each other. Major inconsistencies, which can affect the resolution of important coastal issues should be identified. It is the eventual task of the courts and legislators to rectify inconsistencies through statutory construction and eventual amendments of appropriate laws. A review of mandates, organizational structure and functions of institutions, often reveal gross overlaps in mandates or major gaps in responsibilities, all of which result in jurisdictional conflicts and inter-agency competition. The analyses should aim to guide coastal managers in identifying key institutions that together will provide an optimal mix of mandates and functions to carry out an effective ICM programme.

4.10   Social and Cultural Analyses

Public perceptions about the past, current and future status of the coastal environment and its resources, and how and why they should be managed are invaluable in developing strategies for a coastal management programme. While not expressed in formal instruments such as laws and institutions, perceptions, aspirations and world views directly influence how a society manages its natural resources.   Studies might explain, for example, how certain communities in the Philippines can tolerate the use of blasting as a fishing technique while their laws provide for the death sentence of wilful violators. In such cases, successful enforcement of the law will hinge on the evolution of unequivocal social sanctions, a process which can be facilitated by a culturally sensitive public education programme.

4.11   Management Control Measures

Scientists can provide advice to environmental managers that will help them in selecting and implementing measures to control activities that harm coastal environments. Both natural and social scientists may be required to demonstrate the need for controls. Social scientists, in particular, may be helpful in advising on the optimum choice of control method.

There are advantages in control measures that rely on the voluntary cooperation of those whose activities may affect the condition of coastal resources. For example, voluntary agreements on limiting contaminant discharges to the sea, or limiting rights to exploit resources for particular purposes are often more effective than mandatory (i.e., legislative, regulatory) controls. Such agreements can be accompanied by funding or other incentives to put conservation measures in place. Nearly half of the funds in the Chesapeake Bay programme (see Annex 1), for example, are provided on a cost-sharing basis to farmers to reduce nutrient loadings.

It is often necessary to supplement voluntary arrangements with control measures enforceable under law through permits or other means. For example, areas on land or at sea may be designated, or **zoned**, for almost any purpose that benefits society, the environment or both. In the marine environment, typical applications of zoning include the designation of areas exclusively for aquaculture or watersports, or the designation of specially-protected areas, usually with restricted access, that provide a refuge for important species or that support critical stages in their life-cycles e.g., spawning grounds.

It should be noted that in some societies over-reliance on regulatory authorities can actually reduce public support, because such measures are assumed to be ineffective and subject to corruption. In other places, the public may demand new laws or enforceable regulations; scientists play an important role in setting standards and limits under such approaches. A successful ICM programme, therefore, works constantly to maintain and build its constituencies and to assure selection of the best possible suite of control measures for the implementation of its policies and plans. This is the essence of adaptive management.

4.12   Public Education

Much can be done to advance the overall goal of ICM through public education initiatives organized jointly by managers and scientists. Thus, part of the ICM strategy should be a carefully structured educational campaign, involving, for example, a series of displays, newsletters, lectures and debates, to convey the principles of resource management and related scientific techniques to the widest possible audience. A key topic to address is the concept of **sustainable use** of resources which needs to be explained in both environmental and socio-economic terms. Scientists can make valuable contributions by preparing, in simple terms, information on the characteristics of local ecosystems, indices of productivity and the factors that cause productivity to vary, the inter-dependencies between different ecosystem components, the sensitivities of certain habitats and species and some of the more common hazards to which they are exposed.

Another key focus for public education is the elucidation of forces driving the issues with which the ICM programme is dealing. The processes at work, both social and environmental, should be explained as well as the implications of current trends and conditions.

Scientists should also outline studies of an area and consider possibilities for involving local people in systematic data gathering exercises that, apart from initial instruction, require little scientific knowledge or equipment (e.g., oil spill incidents, occurrence of discarded nets, beach litter, stranded marine mammals, red tide outbreaks, etc.). The profiles of each region developed in the early years of the Ecuador ICM programme *(see Annex 3)*, are an example of issue identification through public involvement.

5. FACTORS AFFECTING THE CONTRIBUTIONS OF SCIENCE AS REVEALED BY THE CASE STUDIES

5.1 Introduction

In preparing this report, GESAMP had the benefit of four detailed case studies. The cases were selected from mature programmes that had either completed or are engaged in the implementation stage. Two cases are from developed nations and two are from developing nations. They represent a range of ecological, geographical, social and cultural environments. These included the Chesapeake Bay Program in the United States, the Great Barrier Reef World Heritage Area Programme in Australia, the Coastal Resources Management Programme in Ecuador, and the Lingayen Gulf and Bolinao Projects in the Philippines. While each case study provides unique insights into the development and operation of ICM, there are also common elements in these real-world experiences. Many of these common elements are related to the role of science in ICM. Synopses of the four case studies are presented in Annexes 1-4.

5.1.1 The Settings

Table III summarizes key social, economic and physical indicators of the national and regional settings for the four case studies. In contrast to the two developed countries, in the Philippines and Ecuador high proportions of the population live in poverty and are directly dependent upon the local resource base for their food and their livelihood. Efforts to conserve or restore ecosystem quality in these areas must respond to the needs of such people. In the case of the Australian and United States examples, the interests which must be dealt with include well-organized agricultural and fisheries industries, as well as organized environmental protection groups. In all four cases, public involvement is critical to success. Also critically important from a resource management perspective are differences in the complexity and layers of existing governance structures, as well as existing governance processes, at both the local and national levels.

In the Philippines and Ecuador, the capacity of institutions at the community and national levels to effectively manage coastal resources was very limited when the programmes were initiated in the mid eighties. Existing regulations were frequently ignored and constituencies representing some sectoral and public interests in ICM did not yet exist. In Australia and the United States, the problem was a large number of overlapping and uncoordinated government agencies which prevented unified action toward commonly held goals. In such situations, a prerequisite for effective resource management is to create within the public service efficient structures and procedures for the administration of environmental regulations and programmes. An efficient governance infrastructure and well established constituencies are preconditions for an ICM programme and also for the enforcement of any zoning scheme, environmental regulations or user agreements.

Table III

Socio-economic Settings of the four Case Studies

PARAMETER	Chesapeake Bay (USA)	Great Barrier Reef (Australia)	Coastal Region (Ecuador)	Lingayen Gulf (Philippines)
1 Gross Domestic Product (USD per capita)	25,818 (1994)	16,700 (1992)	1,062 (1990)	2,660 (1995)
2 Coastline (km)	9,020	2,200	1,256	160
3 Coastal population	14,200,000	400,000	4,700,000	799,000
4 Fishers, incl. part-time	28,500	6,500	30,000	12,500
5 Annual fisheries production (t)	307,900	12,700	390,000	14,400

Note: Items 2 through 5 apply to the areas considered by the case studies; in the USA, the Chesapeake watershed; in Australia the reef and its associated watersheds; in Ecuador the four mainland coastal provinces; and in the Philippines the Lingayen Gulf and its associated coastal municipalities. The length of the coastline of Chesapeake Bay includes all indentations and island shorelines; the estimates of shoreline length for the other sites are measured as straight line distances.

5.1.2 Programme Maturity

Section 3 describes the process by which ICM programmes evolve. The successful completion of a five-stage cycle, beginning with issue identification and ending with an evaluation of programme performance, can be termed a **generation** of a programme. More mature programmes represented by those of the Great Barrier Reef and Chesapeake Bay, have progressed into second or third generations while the Bolinao-Lingayen Gulf and Ecuador programmes have yet to complete the implementation stage of an initial generation. Experience has shown that programmes are well served when learning cycles, in which small-scale pilot projects are used to test the feasibility of implementation, are completed as frequently as possible. The Ecuador programme refers to these as "practical exercises in implementation."

5.2   Analysis of the Case Studies

Analysis of the four case studies shows remarkable agreement among them all regarding the roles, limitations and responsibilities of science in ICM. These findings are organized below into those that are applicable to ICM generally, and those of greatest value to each of the five stages of the ICM process defined in Chapter 3.

5.2.1   ICM Process

Throughout all stages of the ICM process, GESAMP finds that:

- Competent management of a complex ecosystem subject to significant human pressure cannot occur in the absence of science. The natural sciences are vital to understanding the functioning of the ecosystem and the social sciences are essential to comprehending why humans behave in ways that cause ecological problems and can contribute to their solution.

- Science in support of ICM must be undertaken within a structure for solving problems. GESAMP believes that related research and analysis benefit from this strong issue-driven approach. In the absence of special arrangements to provide incentives to the scientific community to focus its research on management issues, scientists may not address directly the questions for which managers need answers.

- Scientists and managers must work together continuously if science is to be relevant and applied to management decisions. The two professions speak different languages, have different perspectives and imperatives, and approach the solution of problems in different ways. They have to learn to work together effectively, for instance in posing management-relevant questions in ways that allow them to be addressed by science.

- There must be recognition of, and mechanisms to deal with, the tensions between scientists and managers, in terms of management's needs to establish clear numerical standards, targets and goals and take action in a situation of scientific uncertainty.

- ICM programmes should deliberately use various means to achieve cooperation and consensus among scientists and managers, ranging from small informal working groups to formal advisory or decision-making committees.

- Managers and scientists must also work together to achieve community support, minimizing the creation of conflict and enmity and maximizing opportunities to identify common interests. The generation of a commitment to a team approach is necessary for real cooperation.

- Community groups must be involved in the design, conduct and interpretation of research that has the potential to lead to management decisions that seriously affect them. Otherwise they are likely to deny the validity or implications of the research results and oppose strongly the decisions based on them.

- Having nearby scientific institutions involved is helpful to achieving these objectives in two ways. The scientists in such institutions are likely to be familiar with the historical and social roots of conflicts and may therefore be able to deal with them, and physical proximity facilitates meetings and joint effort. While there have been great developments in modern electronic communication, nothing is as effective in solving complex problems as a group of people meeting and working together.

- There must be realistic and specific research objectives and time-frames to support a research effort that will successfully answer management questions. The mere description of a management problem followed by the generation of a research programme that is generally relevant to that problem is unlikely to lead to scientific results that can be applied directly to the solution of the problem.

- Science funded by the ICM programme should be subject to peer review; it is especially important for competitive proposals to be reviewed before funding decisions are made, so that scientific cooperation is not jeopardised by suspicions of unfairness.

- While each geographic area is to some extent unique, there is a great deal of scientific knowledge relevant to ICM to build on and to borrow from. Often, special research is not necessary to answer management questions—they can be answered by the application of existing scientific knowledge.

- It must be kept in mind that, within the time available, sometimes science and technology will have only marginal impact on understanding and managing systems as complex as those which are the focus of most ICM efforts; nevertheless, pragmatic management decisions often have to be made.

- Simple and inexpensive technology can often effectively meet the needs of scientists and managers. Furthermore, complex technology can absorb the time and resources of the ICM team while conferring only marginal benefits. Technology should never be applied for its own sake.

- Response to scientific knowledge varies among nations and cultures. The presentation and application of such knowledge must be sensitive to the local culture.

5.2.2   Preparatory Stages

With respect to the **Issue Identification and Assessment and Programme Preparation Stages** of ICM, GESAMP recommends consideration of the following additional findings from the Case Studies:

- The most useful input of science at the first stage is to define management issues, why they are problems, and how they should be addressed, e.g., data to show overfished status of resources. The first task of natural scientists is to supply the objective data to support or challenge perceptions of resource depletion or degradation.

- In these early years, a key role of science is to isolate the causes of the problem and help eradicate misconceptions and prejudices, so that management effort can focus on the real causes.

- There are multiple benefits from emphasizing use of existing data rather than new studies in the first generation—they can provide analytical focus, encourage the involvement of indigenous expertise, allow the definition of trends, provide cost savings and help to target research and management effort. Often, these data already exist and do not have to be generated from new research projects.

- In developing countries, an over-reliance on scientists from developed countries may delay or prevent the establishment of viable ICM programmes, creates high costs and often provides results of limited value due to absence of an acculturation process. The better approach is to take the time to develop an indigenous cadre of scientists, so that the application of science will continue whether or not foreign scientists are involved.

- Indigenous scientific capacity can be built through inter-institutional working groups, mentoring with scientists of international stature, focusing on questions of direct relevance to ICM, and improving the quality and reliability of local work.

- External scientific experience is most useful only after a social and institutional context for ICM is established, because science is never applied in a social vacuum and it is not applied effectively unless its application is integrated into the decision-making process.

- In the early stages, many technology support systems, e.g., GIS and remote sensing, may be available, but actual use will likely be minimal due to time, money, data availability and other constraints. Premature application of sophisticated technology can divert scarce resources from essential activities.

5.2.3 Programme Adoption

As ICM programmes move toward the **Formal Adoption Stage, GESAMP** believes the following additional lessons from the Case Studies should direct the role of science:

- A portion of the research and monitoring activities carried out within the ICM area should be funded by the ICM programme itself, to provide a material incentive to the scientific community to address directly specific management issues.

- Because in many places the reward system for scientists encourages them to concentrate on research which is not relevant to management, there must be additional incentives to scientists to undertake management-relevant research both within and outside the Programme.

- Baselines and monitoring of natural conditions must be in place prior to the Implementation Stage, recording indicators and parameters that will allow assessment of whether the Programme's objectives are being met.

- Baselines and monitoring that document public perceptions and governance procedures should be in place from the beginning, so that the social sciences can be applied to overcoming social problems and the effectiveness of governance can be assessed and appropriate actions taken.

- Scientists can help bring together the kinds of information that are required by managers and politicians to establish clear, challenging and quantified objectives.

5.2.4 Implementation

As ICM programmes enter the **Implementation Stage, GESAMP recommends** that the following additional lessons from the case studies be taken into account:

- There must be **long-term** working relationships and administrative structures that actively pursue and facilitate scientific inputs into management.

- Managers and scientists, working together, must use monitoring results to adapt management, so that the results of monitoring are not misinterpreted and management actions will have the intended effects.

- When choosing among options for intervention, those backed up by better science should gain better acceptance in most communities, although the need for demonstrating the scientific basis for the preferred intervention in terms that are understood by the particular community should be recognised.

- As ICM programmes mature, the role of science evolves from issue identification into helping to develop the needed technologies, and to understanding results of research and monitoring, feed-back loops and other interrelationships.

5.2.5 Evaluation

The **Evaluation Stage** of ICM is critical, although this stage is often neglected. *GESAMP believes* that in this phase the application of science is essential if the efficacy of the management, research and monitoring programmes is to be evaluated objectively and necessary actions taken to set subsequent generations of the ICM programme on course. The lessons demonstrated by the two mature case studies for this stage re-inforce those described above for Stage 1 (Issue Identification and Assessment) and Stage 2 (Programme Preparation) in Section 3.

6.   REFERENCES

Boelaert-Suominen, S. and C. Cullinan, 1994. Legal and institutional aspects of integrated coastal area management in national legislation. Rome, FAO Legal Office, Development Law Service, 118 p.

Chua, T.-E. and Scura, L.F. (eds), 1992. Integrative framework and methods for coastal area management. ICLARM Conf.Proc., (37):169p

Clark, J.R., 1992. Intregated management of coastal zones. FAO Fish.Tech.Pap., (327):167p

GESAMP (IMO/FAO/UNESCO-IOC/WMO/WHO/IAEA/UN/UNEP Joint Group of Experts on the Scientific Aspects of Marine Pollution), 1980. Marine pollution implications of coastal area development. Rep.Stud.GESAMP, (11):114p

GESAMP (IMO/FAO/UNESCO-IOC/WMO/WHO/IAEA/UN/UNEP Joint Group of Experts on the Scientific Aspects of Marine Pollution), 1990. Report of the twentieth session, Geneva, 7-11 May 1990. Rep. Stud.GESAMP, (41):32 p.

GESAMP (IMO/FAO/UNESCO-IOC/WMO/WHO/IAEA/UN/UNEP Joint Group of Experts on the Scientific Aspects of Marine Pollution), 1991. Coastal modelling. Rep. Stud.GESAMP, (43):187 p.

GESAMP (IMO/FAO/UNESCO-IOC/WMO/WHO/IAEA/UN/UNEP Joint Group of Experts on the Scientific Aspects of Marine Pollution), 1991a. Global strategies for marine environmental protection. Rep.Stud. GESAMP, (45):34 p.

GESAMP (IMO/FAO/UNESCO-IOC/WMO/WHO/IAEA/UN/UNEP Joint Group of Experts on the Scientific Aspects of Marine Environmental Protection), 1994. Guidelines for marine environmental assessment. Rep.Stud. GESAMP, (54):28 p.

Hardin, G., 1968. The tragedy of the commons. Science, (162):1243-8

Intergovernmental Panel on Climate Change (IPCC), 1994. Preparing to meet the coastal challenges of the 21st century. Conference report. World Coast Conference 1993. The Hague, Netherlands, National Institute for Coastal and Marine Management (RIKZ), Coastal Zone Management Centre, 49p, Appendices

Organization for Economic Co-operation and Development (OECD), 1993. Coastal zone management. Integrated policies. Paris, OECD, 125 p.

Pernetta, J.C. and D.L. Elder, 1993. Cross-sectorial, Integrated Coastal Area Planning (CICAP): Guidelines and principles for coastal area development. A Marine Conservation and Development Report. Gland, Switzerland, IUCN, 63 p.

United Nations, 1993. Agenda 21, a blueprint for action for global sustainable development into the 21st century. In Agenda 21: Programme of action for sustainable development, Rio declaration on environment and development, Statement of forest principles. The final text of agreements negotiated by Governments at the United Nations Conference on environment and development (UNCED), 3-14 June 1992, Rio de Janeiro, Brazil. New York, United Nations Department of Public Information, pp 13-288.

United Nations Environment Programme (UNEP), 1995. Guidelines for integrated management of coastal and marine areas - with special reference to the Mediterranean basin. UNEP Reg. Seas Rep.Stud., (161):80p

World Bank, 1993. The Noordwijk guidelines for integrated coastal zone management. Paper presented at the World Coast Conference, 1-5 November 1993, Noordwijk, The Netherlands. Washington DC, The World Bank, Environment Department, 21 p.

Annex 1

CASE STUDY 1 - THE CHESAPEAKE BAY PROGRAMME, U.S.A

by William Matuszeski

Chesapeake Bay Programme Office, U.S. Environmental Protection Agency, Annapolis MD 21403, USA

1. BRIEF DESCRIPTION OF THE CONTEXT FOR THE PROGRAMME

Comprehensive coastal management efforts in the United States date from the sixties, when California and a number of other coastal states began to establish state-level agencies to deal with the complex of coastal problems emanating primarily from development pressure. The effort was given a major boost in 1972, with enactment of the Federal Coastal Zone Management Act, which provided funding, structure and other incentives for states to develop for Federal approval so-called comprehensive programmes to deal with land and water uses in a loosely defined coastal zone. By 1979, over two dozen of the 35 coastal states and territories had approved coastal zone management (CZM) programmes covering over 90 percent of the U.S. coast. While these programmes effectively focused attention on the most obvious management needs in littoral areas, they had a number of short-comings: they tended to focus nearly all the effort at the state government level (although a few exceptional programmes such as those in Connecticut and North Carolina made substantial progress bringing in local government); they tended to avoid dealing with areas where there were competing state and Federal bureaucracies (e.g., fisheries management), and they seldom moved upstream enough to capture activities (e.g., agricultural runoff) which were impacting the immediate coastal areas.

Soon after the most active period of coastal zone management programme development, the National Estuary Programme (NEP) was developed. Eventually, over twenty of these were designated by the Federal government and underwent a five year planning process to develop Comprehensive Conservation and Management Plans (CCMPs). They were less focused on **enforceable state oversight** than the CZM programmes but brought more community interest into their design. Their biggest disadvantage was that Federal funding decreased once the CCMP was approved, unlike the CZM programmes, where implementation funds were substantially greater than funding at the planning stage. Integration of CZM and NEP programmes has been made difficult by hostility between Federal oversight agencies (the Commerce Department (NOAA) and Environmental Protection Agency (EPA)), a problem sometimes duplicated at the state level. On the other hand, both programmes have probably benefited from the competition for grass roots support.

The Chesapeake Bay Programme has been able to draw upon some of the best features of both of these programmes. It is a long-term effort based on strong science and with broad public support; this support derives in part from public concern about the Bay, but also from important efforts to make the public part of the restoration through involvement. This, in turn, has resulted in solid funding from the Congress and state legislatures.

1.1 Salient Characteristics of Chesapeake Bay

Chesapeake Bay has a number of unique features. It is the largest estuary in the United States and one of the most productive on earth. Its fisheries alone exceed USD 1 thousand million in value each year, and it creates property values and supports immense recreational investments that serve the local population and are the basis of a major tourist industry.

The Bay is essentially the drowned valley of the Susquehanna River, which today enters the Bay two hundred miles north of the Bay's outlet to the Atlantic Ocean. The two salient characteristics of the Bay are its shallowness and its extensive watershed or drainage area. These are at the same time the reason for its rich productivity and the cause of great challenges to restore its health. The Chesapeake averages about 7m depth, (10% is <1m and 20% <2m). This shallowness creates conditions whereby light penetrates to the bottom, allowing the growth of underwater grasses and other living resources which provide excellent habitat for shellfish and the early stages of finfish. The Bay produces half the blue crab harvest of the Nation in a good year, and it provides over ninety percent of U.S. spawning habitat for the rockfish, or striped bass.

The watershed comprises 166,000 km^2 and includes all or parts of six states and the District of Columbia. While the upper parts of the watershed are primarily forested, the population (14 million) is heavily settled near the Bay and its tidal rivers, and includes the Washington and Baltimore metropolitan areas. These areas also support some of the highest concentrations of livestock and intensive cropping in the United States. Any success in restoring the Bay requires dealing with the diverse sources of pollutants from these highly developed areas.

The unique nature of the Chesapeake system becomes apparent when you combine the two salient characteristics—the shallowness and the extensive watershed—into a ratio of the area of the watershed to the volume of water in the Bay. The result is 2700:1, that is, for every 2,700 km^2 of watershed, there is 1 km^3 of water in the Bay, which is nearly ten times the next closest body of water (the Bay of Bothnia, at 327:1). This also explains why so much of the effort of the Chesapeake Bay Programme is directed at activities on the lands of the watershed and in the streams and rivers that feed into the Bay.

1.2   Wealth of Society

The United States is a developed nation with a per capita domestic gross product approaching USD 26,000, and a relatively low unemployment rate. Some of the wealthiest counties in the Nation are located in the Baltimore and Washington Metropolitan Areas, within the watershed. Sixty percent of the watershed is forested; while nearly all is second growth, the revegetated forests today comprise one of the most extensive stands of mixed hardwoods on earth. There are intensely managed agricultural areas, including large traditional Amish and Mennonite communities who use no electricity or motors. These agricultural areas are the most productive non-irrigated farmlands in the United States. The fisheries of the Chesapeake Bay provide the main source of income for nearly 15,000 watermen, and an equal number are licensed commercial watermen who fish on a part-time basis. Most economic benefits derive from the recreational opportunities and tourism generated by the Bay.

1.3   Existing Governance Structure

Although the Programme is clearly dominated by a state-level "culture", it is well integrated at all levels, with local governments being the most recent to be brought into the management structure. While EPA is the lead Federal agency under the Chesapeake Bay Agreement, there is a high level of cooperation and buy-in by all Federal agencies, manifested by a 1994 Agreement on Ecosystem Management of the Chesapeake Bay, signed at the highest level by 26 Federal agencies. And while the Programme is built on the assumption that environmental regulatory requirements will be met, it is itself consensus-driven. The governing board, or Executive Council, is comprised of the elected Governors of Maryland, Virginia and Pennsylvania, the Mayor of the District of Columbia, the Administrator of the EPA representing the Federal Government and the Chair of the Chesapeake Bay Commission representing the State legislatures. Support of the Programme derives in part from the participation of these high level political leaders.

2.   DESCRIPTION OF THE PROGRAMME TO ESTABLISH ICM

Serious deterioration in Bay water quality and fisheries began to receive widespread attention in the sixties. It was thought that sewage plant improvements would take care of the problem but conditions continued to worsen, with classic eutrophic conditions occurring each summer. Major damage to grasses, shellfish and other living resources resulted from Hurricane Agnes, which brought record amounts of sediment and nutrients into the Bay in June of 1972 and demonstrated how weakened the systems of the Bay had become.

As a result of action by key members of Congress, the EPA was funded with a total of USD 27 million to undertake a five-year study of the Bay between 1978-83. There was strong disagreement in the scientific community over the causes of the Bay's problems. Thermal pollution, toxins, oil spills, dredging and nutrients, all had their advocates. The results identified major concerns, including toxins, declines in grasses, wetlands alteration, shoreline erosion, hydrologic modification, fisheries changes, shellfish bed closures, dredging and shipping. However, a scientific consensus emerged that nutrients were at the heart of many of the problems. Yet, there was political resistance to identifying nitrogen (in addition to phosphorous) as a nutrient to be controlled, because phosphorus removal from sewage treatment plants was considerably cheaper than removing nitrogen. In fact, there was some doubt that nitrogen removal technology even existed. In addition, the focus was still almost exclusively on point sources rather than on non-point sources.

The first Chesapeake Bay Agreement was signed by the states and the Federal Government in 1983. A monitoring programme was started, and steps to improve the water quality of the Bay were taken. While improvements in the rivers could be attributed to phosphorous removal from treatment plants, scientific consensus continued to press that the Bay required that nitrogen levels be substantially reduced.

By 1987, agreement was reached on a broader effort for the clean-up, including both phosphorous and nitrogen controls and looking at the full range of point and nonpoint sources. A new set of commitments was prepared for the principals to endorse. This 1987 Chesapeake Bay Agreement set forth a comprehensive list of 29 commitments to action organised into six areas: living resources, water quality, population growth and development, public education, public access and governance. Scientists had played a major role in shifting the paradigm to target nitrogen as the key nutrient in the more saline portions of the estuary.

3.   GOALS OF THE PROGRAMME

The Agreement had two critical goals. First, that the best ultimate measure of the recovery of the Chesapeake would be the productivity, diversity and abundance of its living resources; this made it clear for the first time that the Programme agreed with the public that water quality was important, not for its own sake but because of how it affects the health of the fish, shellfish, grasses and other living things in the Bay. This led the Bay Programme to much greater involvement with fisheries management, ecological feedback loops in the Bay and work to improve habitat.

The second goal was to reduce nutrients in the Bay. Yet, after all the years of study, the scientists could not agree on a quantifiable goal. The Governors and other members of the Chesapeake Executive Council listened to the scientific evidence and set the goal of a 40% reduction from controllable sources of both phosphorus and nitrogen to the Bay by 2000, using 1985 as the base year. To provide opportunity to further develop models and other information sources, this goal was to be re-evaluated and confirmed or changed after four years.

With this clear goal and direction, public support for the clean-up effort grew, and the role of science began to change to one of supporting the goals and finding the specific technical solutions called for by the Agreement. This included finding ways to remove nitrogen efficiently from sewage treatment plants, new management practices to reduce loadings from fertilizer, manure and other agricultural sources, improved understanding of nutrient stressors on living resources and improved modelling. A substantial percentage of all available funds were directed to cost-sharing programmes for farmers. More research was put into nitrogen removal and agricultural technologies, and limited funds were directed at better understanding the effects of toxins. As the Program grew more operational, a healthy tension developed between scientists and managers; the managers questioned the relevance and timeliness of research being supported by the Program and the scientists questioned the integrity of a management process that does not seek out the very latest scientific results.

4.   MAJOR ISSUES ADDRESSED BY THE PROGRAMME

The major issues are nutrients, toxins and fisheries. All sources of nitrogen and phosphorous are being estimated and tracked, including some which were not originally included in the definition of "controllable". Airborne nitrogen is the source of about 27% of the total entering the Bay, about 9% directly to tidal waters, and about 18% to the watershed and thence delivered to the Bay off the land. Interactive airshed, watershed and 3-D models are in advanced stages of development. Models are being used to estimate the role of living resources in the health of the Bay, including feedback loops from fish and grasses, both of which remove nitrogen from the water. A toxins strategy has also been developed, with major emphasis on pollution prevention actions.

In relation to toxins, new knowledge of possible long-range transport of toxins from the industrial areas of the Ohio Valley and the lower Great Lakes has been obtained. Although many of the toxins of greatest concern are agricultural chemicals, the prime focus of agricultural controls has been on nutrient reductions.

There is less scientific consensus on the relative importance of improved fisheries management to recovery of the living resources. Fisheries science has been a relatively isolated field until recently, so interactions between fisheries scientists and those interested in toxins and nutrients have not been so easy to establish. Also, the assessment of the success and failures of fisheries management itself has only recently been made. The role of the living resources takes on importance not only as a measure of progress but—perhaps more significantly—as a contributor to the cleanup through assuring healthy levels of key nutrient grazers and filterers throughout the ecosystem.

5.   CONTRIBUTION OF SCIENCE

The focus of the first five years (1978-83) of the Bay restoration effort was to isolate the causes of the problems of the Bay. Scientists then played a key role in shifting the paradigms to accept nutrients, especially nitrogen, as the culprits. That done, the next stage was to help develop necessary technologies to deal with them; key among these were biological nutrient removal from sewage treatment plants and management practices for agricultural and urban areas, with particular emphasis on stormwater runoff.

Scientists have been able to assist in bringing together the kinds of information that have allowed clear numerical goals to be established. There are over a dozen such goals set in the Bay Programme, and they are considered a key element in the Programme's need to inform the public of progress and to be able to measure success. The nutrient goal, which is to reduce loadings of nitrogen and phosphorus to the mainstream of the Bay by 40 percent by 2000, using 1985 as the base year, is the most well known. But there are also goals to increase underwater grassbeds (140 km^2 in 1984) to 460 km^2 by 2004, to open up over 2,100 kilometres of streams to anadromous fish by 2005 and to cut toxic releases from Federal facilities by 75 percent between 1994 and 2000. Scientists have had difficulty coming up with such goals themselves; there is always a desire to know more and to avoid specific numerical goals. But scientists can provide a level of confidence to politicians so that they can set the goals with a degree of comfort. And goals with political buy-in have a greater chance of being taken seriously. So, the role of science works well in this context.

Scientific input to the 1992 nutrient revaluation was critical. The fully updated models were used to test the 40 percent reduction goal and alternatives to determine impacts. This work confirmed the essential validity of the goal by showing the impact on oxygen in the Bay. At the same time, associated analyses pointed out the need to focus attention on the individual tributary systems, since reaching the reduction goal would require different sets of actions in each. In its 1992 Amendments to the 1987 Agreement, the Chesapeake Executive Council called for allocations of specific loads of phosphorous and nitrogen to each of the ten major tributary systems of the Bay, and for development with public participation of specific strategies showing how each tributary would achieve the 40 percent reduction by 2000 and hold at that level thereafter.

Scientists are good integrators. As a programme like the Chesapeake Bay Programme matures and moves beyond water quality issues to deal with land activities and airsheds and living resources, the scientific community is helpful in putting the pieces out on the table. This has been especially useful in tracking the full range of nutrient sources, the scope of toxins assessments and the inter-relationships of the water quality parameters and living resources in the Bay. In all this, there is a certain tension, with managers fearful that some scientist will naively undo twenty years of consensus building by coming up with some new theory of what is wrong, and scientists fearful that managers will refuse to correct course when it is clearly shown to be needed.

There is applied science related to the Bay and to its restoration being funded outside the Programme. This includes much university research and support by such funding sources as the Sea Grant Programme. It also includes fisheries stock assessment and related research. These are coordinated through interagency groups or, in the case of the universities, through the Chesapeake Research Consortium (CRC). The CRC maintains a small staff; it seeks to coordinate applied research of all the major universities in the watershed, operates a Chesapeake Bay Fellows Programme that supplies key support staff to the Bay Programme Office and staffs the Bay Programme's Scientific and Technical Advisory Committee.

Science is funded through the Bay Programme itself. Over two million dollars per year goes into Bay modelling efforts, and a million more into monitoring. These include funds for independent oversight of Programme activities. In addition, competitive proposals are requested in such areas as toxic assessment and nonpoint source controls. These are generally focused on finding the most cost-effective technologies to reduce loadings to the Bay.

The Scientific and Technical Advisory Committee (STAC) plays a number of important roles. It is comprised of top scientists from throughout the watershed; they are appointed through a variety of means, including some by each state Governor and some by EPA. They also have the authority to augment their membership by selecting members to cover otherwise unrepresented disciplines. The STAC establishes peer review systems for all Bay-funded competitive research, reviews and comments on all proposed budget items, holds symposia and carries out technical reviews of key scientific issues.

6. LESSONS LEARNED ON THE ROLE OF SCIENCE

First, science is essential at the outset to make sure the right issues are being dealt with. Sometimes public opinion will drive efforts in the wrong direction. Science must be willing to "take on" these opinions and turn them. Otherwise, either the effort is unsuccessful or it does not have the requisite public support to get the needed resources.

Second, scientific consensus on the nature of the problem is essential; sometimes this requires major paradigm shifts. There is little hope for integrated coastal management if there is disagreement among the scientists. No political leaders want to support investments in restoration that some scientists are saying will not work. Of course, there is an occasional individual who can't be brought on; but general agreement on the technical fixes is a key to support. And you better not be wrong; that is why paradigm shifts are often needed at this stage.

Third, management efforts must begin as soon as the issues are identified. Integrated coastal management is not an effort where you can figure it all out at the outset; it is more like peeling an onion. As you reach each layer, you learn new things to help you understand the next layer. There is never enough time and money available at one time to act otherwise. In the case of the Chesapeake, it took from 1978-83 to nail down the nutrient issue, from 1983-87 to agree we had to go after nitrogen, until 1992 to begin to understand the interactions and feedback loops of the living resources, and until 1994 to begin to look at air deposition as a major source of nitrogen. If we had tried to figure all this out in 1983, or even had taken until 1987, we would have long before lost all public support.

Fourth, science in support of integrated coastal management must be management-driven within a structure for solving problems. There are simply not enough resources to do otherwise. The five years and USD 27 million EPA had to study the problems of the Chesapeake is a luxury of the past; even if that were not so, there were many at the time who thought much of the money was thrown at the problem and wasted. It took the Programme years to overcome its reputation as a scientific spendthrift. Once there is agreement on the issues to be addressed, science in an ICM context is most useful when directed at the cost effectiveness of specific options, the feed-back loops of the natural systems and other issues of inter-relationships. Using monitoring, good quality assurance of data, modelling and environmental indicators of results and progress, science can then signal when it is time to peel off another layer of the onion.

Fifth, there is a lot of knowledge to build on; it is not necessary to reinvent the wheel. Much of the technology used in the Chesapeake to remove nitrogen from sewage treatment plants was borrowed from South Africa. Similarly, there is little need to study the causes of eutrophication in highly stratified estuaries with populated watersheds in temperate zones; we can already tell you that, likely, the problem is phosphorous-driven in fresh water areas and nitrogen-driven in more saline waters. We can also tell you where to look for sources of the nutrients, and how likely it is that, regardless of what you think now, you will have to deal with both phosphorous and nitrogen. We spent decades learning this; but very little time is spent spreading the word. And scientists will always say "It's different here". May be it is, but you can find that out down the line.

Finally, success is measured against goals, and science must be willing to give politicians the level of confidence needed to have them set goals that are clear, measurable and challenging.

It is not for the scientists themselves to set such goals; even if they could reach agreement (a very big "if"), the goals would lack the necessary political endorsement. Nor are bureaucrats the right ones to set the goals; they tend to be too timid and too much in favour of the vague platitude. So, how can the political leaders get the advice they need to set the right goals? They have to rely on scientists for the courage to overcome the advice they are getting from their bureaucratic underlings to avoid specifics and timeframes. This is an odd but essential role for science. And it is not that difficult, since most political leaders have good instincts for what the public wants and are willing to bear for the common good.

Annex 2

CASE STUDY 2 - THE GREAT BARRIER REEF, AUSTRALIA

by Graeme Kelleher[1]

Graeme Kelleher & Associates Pty Ltd, POBox 272, Jamison ACT 2614, Australia

1. SALIENT CHARACTERISTICS OF AUSTRALIA AND THE GREAT BARRIER REEF

1.1 Conditions in Australia

The coastal strip and sea are very important for Australians. A quarter of the population lives within three kilometres of the sea and two-thirds reside in coastal towns and cities. Australia's 200 nautical mile Exclusive Economic Zone (EEZ) is far larger than the land area and is one of the largest EEZs in the world. It is difficult to assess simply and accurately the general condition or environmental health of Australia's marine environment owing to its vast size, its great diversity, the range of issues affecting it and the large gaps in scientific knowledge (Zann, 1995).

On the basis of the existing limited information, and in comparison with both neighbouring countries and equally developed countries in the northern hemisphere, the state of Australia's marine environment is rated as **generally good**, but with many important qualifiers. The condition of specific areas ranges from virtually pristine (natural or unspoilt) off remote, undeveloped areas to locally poor off highly developed urbanised, industrialised and intensively farmed areas.

Australia's population is highly concentrated in coastal cities in the south-east and south-west. Here, the state of the adjacent marine environment may be locally poor. So, while the state of Australia's marine environment is on average good, the state of the marine environment near where the urban Australian lives is often **'not good'**.

The top five concerns relating to Australia's coastal and marine environment have been identified (Zann, 1995) as:

1. Declining marine and coastal water/sediment quality, particularly as a result of inappropriate catchment land use practices

2. Loss of marine and coastal habitat

3. Unsustainable use of marine and coastal resources

4. Lack of marine science policy and lack of long-term research and monitoring of the marine environment

5. Lack of strategic, integrated planning in the marine and coastal environments

---

[1] former Chairman of the Great Barrier Reef Marine Park Authority

1.2   The Conditions Applying to the Great Barrier Reef

The Great Barrier Reef is unique, and the commitment of the Australian people to its conservation is great. Biologically, the Great Barrier Reef supports one of the most diverse ecosystems known. It has developed over 500,000 years on the northeast Continental Shelf of Australia.

The area extends over approximately 2,200 km along the eastern coast of Queensland, north of Fraser Island in the south (24°30'S) to the latitude of Cape York in the north (10°41'S) and covers an area of 348,700 km^2 on the continental shelf of Australia. It is acknowledged as an area of great natural beauty. The unique environment of the Great Barrier Reef, its size and diversity, have been recognised world-wide and led to its inscription in October 1981 on the UNESCO World Heritage List.

The Great Barrier Reef is not a continuous barrier but a broken maze of coral reefs, some with coral cays. Some 2,900 individual reefs, including 760 fringing reefs, lie within the formally defined area known as the Great Barrier Reef Region. There are some 300 reef islands or cays; 87of them are permanently vegetated. There are about 600 continental or high islands, often with fringing reefs around their margins.

1.3   Wealth of Society and its Dependence on the Great Barrier Reef for Livelihoods

In conventional economic terms, Australia is a wealthy country, its per capita gross national product (GNP) being USD 17,400 in 1993 (Europa World Year Book, 1995). In terms of overall wealth, taking account of environmental values not normally measured by GDP, Australia is considered by the World Bank to be the most wealthy country in the world on a per capita basis (Zagorin, 1995). Its society is not characterised by the extreme disparities in individual or community wealth evident in many other countries, although there has been a trend over the past decade towards greater disparity.

It has been estimated that the value of reef-related activities (on the Reef and on the adjacent mainland) approximates USD 1,100 million per annum, of which USD 750 million are generated by reef-based activities in the Great Barrier Reef World Heritage Area (GBRWHA). Commercial fishing and tourism; recreational pursuits including fishing, diving and camping; traditional fishing; scientific research and shipping all occur within the GBRWHA.

Resort tourism is the largest commercial activity in economic terms (ATIA, 1984).

There is conflict among some groups of people who use the Reef (for example between commercial and recreational fishers). There is also conflict between people who wish to exploit the Reef and those who wish to see it maintained in its pristine state forever. Some uses of parts of the Reef have already reached levels which may fully exploit the productive capacity of the system. Bottom trawling for prawns is an example. Run-off from islands and the mainland contains suspended solids, herbicides, pesticides, nutrients and other materials. The magnitude of their effects on the Reef is not yet known, but intensive and extensive scientific research is proceeding within an integrated research programme, to which the Great Barrier Reef Marine Park Authority contributes substantially, in order to answer these questions unequivocally.

1.4   Existing Governance Structure and its Prior Effectiveness in Successfully Implementing Natural Resource Strategies

Australia has a federal system of government consisting of three levels—the federal (Commonwealth) level, the state or territory level and the local government.

Although the Constitution gives the states and territories primary responsibility for natural resource management on Australian land, the federal Government has the ability to greatly affect such management through its grants scheme, the power of which is magnified by its retaining the sole right to impose taxes.

The situation relating to Australia's seas is different. The federal government has the constitutional responsibility for all the marine areas subject to Australian jurisdiction other than those narrow areas bordering the coast that are defined as internal waters and are within a state or territory. This provision was modified by the governments of Australia in 1979 by extending the powers of each State to the adjacent three nautical mile Territorial Sea and vesting in each State title in respect of the Territorial Sea seabed. These changes were made subject to valid Commonwealth law continuing to prevail over conflicting state law (Commonwealth of Australia, 1980).

It is important for this study to note that, although the legislation which created the Great Barrier Reef Marine Park Authority (the Authority) is federal (Commonwealth) legislation, the Act continues to apply to the whole of the Great Barrier Reef (GBR) Region, including that part of it within the three mile Territorial Sea (Ibid).

Responsibility for various important coastal resource decisions is divided between the three levels of government, with local government having significant powers—particularly in relation to coastal development.

Prior to the enactment in 1975 of the Great Barrier Reef Marine Park Act (the Act), resource management in the GBR Region was further fragmented by management being carried out on an individual resource basis, with quite inadequate coordination between the different management agencies, extending in some cases to open hostility. Thus, decisions were often made in one sector without regard to the effects on other sectors or on the ecosystem as a whole.

Many government enquiries have identified the fragmented and often duplicate responsibilities in the coastal zone as severe impediments to effective planning and management (Resources Assessment Commission, 1992).

2.   DESCRIPTION OF THE PROGRAMME TO ESTABLISH ICM OVER THE GREAT BARRIER REEF WORLD HERITAGE AREA

2.1   Introduction

The Programme and its goals and objectives have developed over a period of 20 years. It was initiated in consequence to the objection of many Australians to proposals to mine coral on the GBR for lime for cement and to drill for oil in what was known as the Reef Province. This led to the federal Parliament passing the Great Barrier Reef Marine Park Act in 1975, with the support of all political parties.

The Act established the three-person Authority and made provision for the establishment of a marine park in the GBR Region, which encompassed all the waters east of the State of Queensland from the northern tip of Cape York Peninsula to a point approximately 2,200 km south and from the low water mark on the mainland to the edge of the continental shelf. The Region also formally encompassed the few islands within those boundaries which were owned by the Commonwealth but omitted the approximately 900 islands which formed part of Queensland and were not owned by the Commonwealth.

The creation and management of the Marine Park can be seen as **Generation One**, or the first Generation of the ICM Programme. This Generation continues and is overlapped by **Generation Two**, or the second Generation of the Programme, which commenced with the listing of the Great Barrier Reef, together with all the islands in the GBR Region, on the World Heritage List in 1981.

2.2 Goals of the Programme

GENERATION ONE

The Act gave no indication of the proportion of the Region that might eventually become part of the GBR Marine Park. Neither did it define a goal other than to establish such a Park. However, it did require that sections of the Park should be zoned and that, in the making of a zoning plan, regard be paid to the following objectives:

"(a)   the conservation of the Great Barrier Reef;

(b)   the regulation of the use of the Great Barrier Reef so as to protect the Great Barrier Reef while allowing the reasonable use of the Great Barrier Reef Region;

(c)   the regulation of activities that exploit the resources of the Great Barrier Reef Region so as to minimize the effect of those activities on the Great Barrier Reef;

(d)   the reservation of some areas of the Great Barrier Reef for its appreciation and enjoyment by the public, and

(e)   the preservation of some areas of the Great Barrier Reef in its natural state undisturbed by man except for the purposes of scientific research."

The Great Barrier Reef Marine Park Authority has derived a primary goal and a set of aims from the provisions of the Act and recognition of the political, legal, economic, sociological and ecological environment in which it operates.

The primary goal is "To provide for the protection, wise use, understanding and enjoyment of the Great Barrier Reef in perpetuity through the development and care of the Great Barrier Reef Marine Park."

The set of aims derived, which are subordinate to the primary goal and must be read in conjunction with it and with each other, are:

"(1)   To protect the natural qualities of the Great Barrier Reef, while providing for reasonable use of the Reef Region;

(2)   To involve the community meaningfully in the care and development of the Marine Park;

(3)   To achieve competence and fairness in the care and development of the Marine Park through the conduct of research, and the deliberate acquisition, use and dissemination of relevant information from research and other sources;

(4)   To provide for economic development consistent with meeting the goal and other aims of the Authority;

(5)   To achieve management of the Marine Park primarily through the community's commitment to the protection of the Great Barrier Reef and its understanding and acceptance of the provisions of zoning, regulations and management practices;

(6)   To minimize the costs of caring for and developing the Marine Park consistent with meeting the goal and other aims of the Authority;

(7)   To minimize regulation of, and interference in, human activities consistent with meeting the goal and other aims of the Authority;

(8)   To achieve its goal and other aims by employing people of high calibre, assisting them to reach their full potential, providing a rewarding, useful and caring work environment and encouraging them to pursue relevant training and development opportunities;

(9)   To make the Authority expertise available nationally and internationally;

(10)   To adapt actively the Marine Park and the operations of the Authority to changing circumstances."

Taken together, this goal and these aims constitute a set of objectives and strategies for Generation One or the first generation of the ICM Programme.

GENERATION TWO

The second generation in the ICM iterative process commenced theoretically with the World Heritage listing of the GBR in 1981, in the sense that the whole coastal ecosystem was formally recognized as an entity, including all the lands, seabed and waters within the outer boundaries of the GBR Region. However, the impetus to undertake the development of a long term strategic plan came from a number of sources about ten years later, in 1990-1991. It revolved around the fact that no regional strategic framework existed because there was no general agreement on the meaning of the term "conservation while allowing reasonable use" which is embodied in the Act.

In consequence:

*   there was a need for a reef-wide (i.e., ecosystem-wide) rather than Park-section view in setting objectives;
*   specific target objectives against which to assess management needed to be set;
*   different agencies might be working with widely differing objectives;
*   the willingness to work together between agencies was not being capitalized on;
*   stakeholders were keen to see targets set, and
*   no single agency had overall jurisdiction.

The goal of Generation Two of the ICM Programme was thus to describe a future vision for the Great Barrier Reef World Heritage Area (GBRWHA) and to determine objectives and strategies which would ensure that this vision was achievable.

The process of agreeing on what the Area should look like in 25 years time produced a unanimous future vision, one that emphasises environmental protection while maintaining sustainable use. The vision described in the Plan which was produced in this second Generation is:

"In the Great Barrier Reef World Heritage Area in 25 years we will have:

*   a healthy environment
*   sustainable multiple use
*   maintenance and enhancement of values
*   integrated management
*   knowledge-based but cautious decision-making in the absence of (specific) knowledge
*   (an) informed, involved, committed community (GBRMPA, 1994)"

Subsequently, the long-term and short-term objectives and strategies to achieve this vision were developed in the areas of integrated planning, conservation, resource use, communication, research and monitoring, Aboriginal and Torres Strait Island, management processes and legislation.

3.   MAJOR ISSUES ADDRESSED BY THE PROGRAMME

3.1   Overview

The major issues dealt with in both Generation One (the GBR Marine Park) and Generation Two (the development of the Strategic Plan for the World Heritage Area) of the Programme are essentially the same. In general terms, they are those listed in the preceding paragraph. What differed between the two generations was the approach taken by the principal coordinating agency (the Authority), the attitude of the participants and the resources devoted to cooperative work.

The linkages between the issues are recognized in the quotation:

"The environment does not exist as a sphere separate from human actions, ambitions and needs." (World Commission on Environment and Development, 1987).

This recognition pervaded all of the Authority's work from the first day, and became progressively more widely expressed, either tacitly or overtly, by all of the more than 60 organisations involved in both generations of the Programme. One of the major functions of the Authority's community education programme was to raise the awareness and acceptance of this principle throughout Australia and beyond.

This case study shows that it is possible in an educated and active society to create a general awareness of the interconnectedness of human concerns and the natural environment and of their elements, provided that there is an agency which is committed to this.

The Great Barrier Reef Marine Park Act was one of the first pieces of legislation in the world to apply the concept of ecologically sustainable development to the management of a large natural area. Real public involvement in all areas of management and decision-making is at the centre of the strategic approach adopted by the Authority to ensuring that human use of the Great Barrier Reef is ecologically sustainable. So far, the approach has been successful—over-exploitation of the Great Barrier Reef has largely been prevented.

The development and application of the GBR World Heritage Strategic Plan is a further significant step in achieving fully integrated coastal management.

3.2   Significance of Each Issue

The following issues deserve specific attention in this case study because they demonstrate relationships or processes which have universal relevance:

- integrated management;
- pollution;
- physical alteration of the seabed or coastline;
- infestations of crown-of thorns starfish (*Acanthaster planci*);
- overfishing;
- introduction of exotic species, and
- climate change.

In the following part of this case study, each of those issues is reviewed briefly in relation to its significance, the objectives and strategies derived to deal with it, the major actions taken and the major outcomes of those actions. For the sake of brevity, the other issues are not dealt with here.

In every case, the objective has been to prevent deterioration of the GBR ecosystem as a consequence of human activity. The strategies used have in each case included that of bringing the critical people and organisations together to define the problem, to derive solutions and to implement them cooperatively. In summary, the principles of integrated management have been applied without exception. Also without exception, the progress that has been achieved in each of these policy areas has exceeded that attained anywhere else in Australia. Evidently, this extraordinary progress has been due to the integration flowing from the existence and actions of the Authority. It can be safely concluded that such integration and its consequent benefits are unlikely to occur anywhere, in the absence of an agency with the explicit functions of achieving integrated planning and management and ecologically sustainable development of a complete ecosystem, i.e., integrated coastal management (ICM).

INTEGRATED MANAGEMENT

As described in the previous section of this case study, ICM has since 1975 been the principal issue dealt with by the Authority, because it is seen to be the single overriding determinant of whether human use of the GBR can be managed so as to be ecologically sustainable.

In a few words, the principal objective of ICM as applied on the GBR has been ecologically-sustainable development. The application of the methods embodied in ICM has been the strategy for achieving this objective. The major actions taken in the programme have been:

- the passage of the GBRMP Act and the establishment of the Authority;

- the development and administration of zoning plans, with a uniquely high degree of public participation;

- the development and application of major, integrated public education and information programmes;

- the development, implementation, interpretation and application of the results of comprehensive, multidisciplinary research programmes, covering the whole World Heritage Area as well as the catchments on the mainland that drain into that Area;

- the development and implementation of a comprehensive, integrated monitoring programme, covering the whole World Heritage Area as well as the catchments on the mainland that drain into that Area;

- the listing of the GBR on the World Heritage List, and

- the bringing together of all the major stakeholders in the GBR to produce the strategic plan for the Area.

The major outcomes of the Programme have been the public acceptance of all of the actions listed above and the successful completion of the strategic plan, involving the unanimous adoption by the more than 60 stakeholder organizations of the plan's vision and 25-year and five-year objectives. In summary, all the major attributes of ICM have been achieved to a significant degree on the GBR.

POLLUTION

The most serious concern for the future of the Great Barrier Reef is deterioration in water quality caused by increases in suspended sediments, nitrogen and phosphorus emanating from the mainland. Reef-building corals are very vulnerable to increases in the levels of these materials. Monitoring has shown that the porosity of some near-shore reefs of the Great Barrier Reef has been increasing over the past few decades due to increase in phosphorus levels. The levels of nitrogen in some parts of the Great Barrier Reef at times exceed significantly the levels which have been shown to cause death in branching corals permanently subjected to them. Protection of the Great Barrier Reef from increasing nutrient levels may be the greatest challenge facing the Authority in the next two decades.

Probably the major issue is run-off from farm land of nutrient- and sediment-enriched water. However, even in this case, high levels of cooperation in research have been achieved between the Authority, farmers' organisations and state Government agencies responsible for primary industry. The experience has been that if cooperation is achieved in carrying out research into a problem, then that cooperation is likely to extend into defining and applying ways of solving the problem.

Over the past few years, the Authority has received major grants from the federal Government to investigate this issue. This has allowed the establishment of a reef-wide monitoring programme and a multi-institutional, inter-disciplinary research programme which aim at developing a complete understanding of the origins of these nutrients and sediments so that corrective action can be taken if necessary.

The objective in relation to land-based sources of pollution has been to limit them to levels which do not cause significant changes in the GBR ecosystem, using as a baseline the general perceived conditions prevailing when the first significant research was carried out on the Reef, namely the research expedition mounted by the Royal Society under the leadership of Sir Maurice Yonge in 1929-1931.

The most important strategy has been the involvement of farmer organizations in cooperative research projects (with the Authority and other organisations) which address this question and to work with them to develop methods of reducing the loss of soil and nutrients from farmland. The commonality of interest of the farmers (in wishing to reduce expenditure on fertiliser and reducing soil erosion) and of the people and organisations who wish to protect the quality of GBR waters has been deliberately identified and used as a means of achieving cooperation (i.e., integration). The action has been to compare nutrient run-off figures for catchments that remain in a state similar to that prevailing before European settlement with those in modified catchments, to determine experimentally and by other means the effects on organisms and biological communities of increased nutrient levels, to determine the causes (origins) of those nutrients and to modify land-use practices to reduce the amount of nutrients entering GBR waters. The outcome has been to reveal that the levels of nutrients entering GBR waters from the mainland have been shown to be about four times higher than those prevailing four or five decades ago (Moss *et al.*, 1992); that increased nutrient levels enhance the porosity of certain corals and reduce their capacity to compete with algae; that most of the increase in nutrients comes from grazing land; and that the levels of erosion and nutrient loss from sugar cropping can be reduced by an order of magnitude by changes in farming practice—changes which have been voluntarily adopted by most sugar farmers on irrigated land on Queensland's coast.

PHYSICAL ALTERATION OF THE SEA-BED OR COASTLINE

The major concern in this subject area on the GBR is the effect of bottom-trawling on the ecosystem. The significance of this issue cannot be assessed in the absence of the results of the long-term research programme presently underway. However, it is rated as one of the most important areas of uncertainty in attaining the objective of ecologically sustainable development of the GBR.

The major action is to carry out a long-term integrated research programme with the aim of defining the ecological effects of bottom trawling. The strategy is to involve the principal stakeholders in this issue. In particular, the fishing industry is a participant in this research programme, so that it will be committed to the results and cannot disclaim them. The insistence of the Authority on this involvement led to a delay of many years in the commencement of the research programme, and it should be emphasised that this insistence was a consequence of many disheartening experiences when research projects were carried out without the involvement of critical stakeholders, resulting invariably in those stakeholders not accepting research results which they perceived as contrary to their interests. The outcome has been the generation of improved trust and cooperation between the management and research agencies and the fishing industry, although the relationship will always be fragile, and individual fishers will deliberately attempt to sabotage or discredit research that might lead to increased control of fishing. Of course, the outcome of this research cannot be predicted.

The issue of coastal development has been mentioned previously. While the mainland coast is outside the jurisdiction of the Authority, there is little doubt that the existence and work of this agency has contributed to the adoption by the Queensland Government of integrated planning as a normal mechanism, most recently reflected in the preparation of a coastal planning bill.

CROWN-OF-THORNS STARFISH

The publicly perceived significance of this issue is highly variable as is the visibility of the phenomenon. At times when infestations occur, there are strident claims by some people, including a few scientists, that the whole GBR is at risk and that millions of dollars should be spent on eradication programmes. When the infestations pass, such demands disappear, but reappear at the commencement of any subsequent population increase.

The objective of the Authority and the federal government is to determine whether human activity increases the severity or frequency of infestations and to act in a way which protects the natural qualities of the GBR. The strategy is to carry out comprehensive research and public education programmes which address this issue while avoiding major interference in what may be a natural phenomenon which contributes to the GBR as an ecosystem.

If it were scientifically demonstrated that human activity is a major factor, then the Authority would react in two ways. First, it would reconsider its policy of limiting control of population numbers to areas of particular scientific or tourist value. Second, it would have to move strongly to ensure that the human activities which were shown to be causative factors were modified.

If, on the other hand, scientific evidence were to accumulate proving that outbreaks have not changed in intensity or frequency since Europeans arrived in Australia, the Authority would maintain its existing policies. The major actions have been to carry out a major research and monitoring programme to interpret the results to the public and to counter by reasoned argument demands for the Authority to treat the starfish as an alien species. The outcome has been the periodic decline and subsequent increase in starfish numbers, without wholesale human interference. This cycle has been parallelled by the degree of public concern and controversy.

4. CONTRIBUTION OF SCIENCES AND SCIENTISTS

"How complex and unexpected are the checks and relations between organic beings, which have to struggle together in the same country!" (Darwin, 1906).

An eminent scientist has said that the Great Barrier Reef is the most complex system in the universe. This statement illustrates two things. First, it recognises that the Great Barrier Reef is highly complex and that the processes that structure and control it are correspondingly complex and difficult to manage. Second, it illustrates that even experienced scientists sometimes make unfounded (or unscientific) statements. In controversial circumstances, these statements can cause great difficulties, not only for science, but also for managers.

This section summarises how science contributes, both positively and negatively, to management of the Great Barrier Reef and gives some examples of the controversy that is generated when scientists make adamantine statements on controversial issues which affect the public.

Adequate knowledge of the baseline or reference ecological characteristics of the Reef is essential in order to monitor the changes brought by human activities. It is also necessary to be able to predict roughly the type and scale of effect likely to be produced by individual activities and combinations of them, so that the intensity and distribution of usages can be controlled—but not over-controlled—in a manner compatible with the conservation of the Reef's natural qualities.

To the maximum extent practicable, the Authority relies on research carried out by specialist research organisations for the scientific information necessary for management. However, its consistent experience has been that the normal framework within which science is conducted in Australia does not provide incentives for scientists to do work which is relevant to management. On the contrary, until recently, there have been strong disincentives provided by the reward system under which university scientists operate in Australia against scientists carrying out management-relevant research.

It would be wrong not to recognise the dramatic changes that have occurred in the management of science in Australia over the last few years. It is clear to us as managers that, contrary to what was true in the past, many more scientists are now becoming interested in carrying out science that is relevant to managers. I think the reason for this is money, which is increasingly being provided for application of science in what have been identified as priority areas.

I would now like to discuss briefly how the scientific community has contributed to controversy and how that controversy has affected competent management in relation to two issues with which this Authority has been concerned. They are: (1) crown-of-thorns starfish and (2) the effects of nutrients on coral reefs.

4.1   The Crown-of-thorns Starfish Controversy

This subject became controversial because practising scientists made statements, apparently unsupported by any significant scientific evidence, that the Great Barrier Reef was doomed. The first such statements were made in the early sixties. Governments were exhorted to spend millions of dollars killing crown-of-thorns starfish because otherwise the Great Barrier Reef would be entirely degraded within ten years and the Queensland coast would be vulnerable to erosion from the great waves of the Pacific Ocean.

It was noticeable and it remains noticeable today, that the scientists who created the hysteria also applied for research funds to allow them to investigate the subject.

As we now know, the first carefully observed crown-of-thorns infestation which started in 1962 or thereabouts, petered out about ten years later. Demands for action and claims that the Great Barrier Reef would be destroyed abated at the same rate as did the crown-of-thorns infestation. They were renewed, using exactly the same expressions, but with a time-shift of 20 years in relation to predictions of destruction of the Reef, when the second carefully-observed infestations started in about 1979. Not only did the biological phenomenon follow the same pattern as the previous episode but the behaviour of some of the scientists was also identical.

I do not want to imply that the Great Barrier Reef is not doomed. That question still remains open and it is vital that we continue research until we can answer the fundamental question of whether human activity increases the frequency or severity of crown-of-thorns starfish outbreaks. This question is vital, because if it is demonstrated to the satisfaction of the majority of scientists and decision-makers that human activity has exacerbated outbreaks, then management agencies, and the Great Barrier Reef Marine Park Authority in particular, would change their present policy of not interfering massively in the phenomenon.

4.2   The Effects of Nutrients on the Great Barrier Reef

There is increasing controversy about whether the inshore parts of the Great Barrier Reef are being degraded because of increases in nitrogen and phosphorus mainly emanating from run-off from the mainland.

A recent manifestation of this controversy appeared in an edition of the scientific journal *Search*. Here, two scientists with extremely opposing views on whether or not there was degradation of inshore reefs on the Great Barrier Reef and whether or not this could be attributed to increased nutrient levels in the waters of the Reef used selective quotations from the same scientific paper prepared by a third party as major arguments for supporting their views. In reading one of these two conflicting papers in isolation, an uninformed person would have been persuaded that the paper being quoted from came to the unequivocal conclusion that parts of the Reef were being degraded by increasing nutrient levels. Another uninformed person, reading the paper presenting the opposite view, would also have believed that the paper being quoted from had come to an unequivocal conclusion. The irony is that the two readers would have derived entirely opposite impressions regarding the conclusions arrived at in the paper from which the quotations were drawn.

The people in the Authority retain an open mind on the subject, while harbouring the suspicion that there has been progressive degradation of inshore reefs of the Great Barrier Reef system over the past few decades and while accepting that, if this is correct, increased nutrient levels in the waters of the Reef have probably contributed to this degradation. As with the crown-of-thorns controversy, we believe that the correct response is to establish an integrated multi-disciplinary research programme aimed at answering the following questions unequivocally:

•   Have parts of the Great Barrier Reef been degraded over the past few decades?

•   If so, can that degradation be attributed to increased nutrient levels?

•   If so, what are the sources of those nutrients, and can human practices be changed so as to reduce nutrient levels in the future?

4.3   Conclusion

If the Great Barrier Reef is to be used in an ecologically sustainable manner in perpetuity in accordance with the Authority's primary goal, then it must be managed on the basis of scientific information.

The amount of funds available to carry out research necessary to achieve this goal is limited. It is essential that the funds be directed towards those areas of research which have the highest likelihood of answering management questions. The two brief examples of the behaviour of the scientific community in situations of controversy have been outlined in order to illustrate firstly, that parts of the scientific community do not act scientifically in promoting hypotheses; secondly, that such actions establish an air of hysteria; and thirdly, that the effect of such hysteria is to divert scarce resources towards solution of the problems causing hysteria and away from other management-relevant questions which may be equally or more important.

It seems to me that such behaviour by scientists confuses the public, throws the profession into disrepute, makes the task of management more difficult and diverts funds and other resources into the hands of the scientists that make the most noise. Some people might think that this is a misuse of science.

On the other hand, it is sometimes argued that anything that raises the visibility of science is good and that one of the most effective ways of doing this is to create controversy deliberately by exaggerating the certainty and severity of scientific conclusions.

Finally, I should conclude by saying that the Great Barrier Reef Marine Park Authority enjoys excellent working relationships with the scientific community generally and depends absolutely on their work for ensuring that the Great Barrier Reef is protected forever.

5.   LESSONS LEARNED ON THE ROLES OF SCIENCE

There has been a great deal of change and development in the relations between scientists and managers (decision makers) since the Authority was created in 1975. Many of these changes have been introduced as a result of painful experiences. It is worthwhile to consider briefly the lessons that we have learned over the past 20 years so that the community can minimize the tendency to repeat the mistakes of the past.

5.1   How it Was

Twenty years ago research and management relating to the Great Barrier Reef were characterised by disintegration. There was disintegration within the research community, between research institutions and within research institutions, and there was a lack of cooperation between and within management institutions. There was little communication between research institutions and management institutions. The intensity of this phenomenon varied but it sometimes extended to hostility between individuals and organisations.

In those days, little was known about the Great Barrier Reef in physical, chemical, biological or ecological terms.

5.2   How it Is

Today the situation is very different. There is cooperation between scientific institutions, between different management institutions and between the scientific and management communities. Recently, the extent of this integration has increased markedly, the most significant evidence of it being the creation of the Cooperative Research Centre for Ecologically Sustainable Development of the Great Barrier Reef (the CRC).

Now, we are less ignorant than we were 20 years ago but we are still far from understanding fully the processes which structure the Great Barrier Reef.

5.3   What have we Learned?

The lessons that those of us who have been here for a long time have learned include:

• Scientists and managers must work together continuously if science is to be relevant to management and if science is to be applied to management decisions. It is not enough for the relationship between the two groups of people to be staccato or occasional.

• Science makes mainly marginal increases in understanding. The results are rarely unequivocal. There is dispute even on fundamental issues — in the case of the Great Barrier Reef on whether nutrients are adversely affecting the Great Barrier Reef or on whether trawling adversely affects either the bottom communities or adjacent coral reef communities.

• Managers must make decisions, whether or not unequivocal scientific information is available. We have learned that managers should base their decisions on:

    - trends rather than states;

- the precautionary principle, so that where there is doubt about the outcome of the matter the decision should err on the side of preventing environmental damage;

- priorities, *i.e.* management effort and scientific effort should be related to the importance of the issues. At present we are far from this.

• Scientists are unlikely to address management issues unless there are incentives provided within the system for them to do so. Experience has shown that the transmission of a proportion of the funds for research through the management agencies will provide such incentive.

• Managers and scientists, working together, must monitor the results of management decisions and adapt management to the results of that monitoring.

• Managers will never be successful without community support. In a democratic society, governments follow community opinion. Therefore managers and scientists must work to achieve community support for decisions which protect the ecology of the Great Barrier Reef.

• Critical stakeholders must be involved in the design, conduct and interpretation of research that has the potential to lead to management decisions that seriously affect them. Otherwise they are likely to deny the validity of the research results and oppose strongly the decisions based on them.

• There are many enemies, both potential and real, in the community. Our mutual efforts will only be successful if we minimize the creation of enemies and maximise the opportunities to identify common interests. A particular example of this is the issue of run-off from the mainland of nutrients and suspended sediments. Farmers are just as interested as are those who care for the Great Barrier Reef in preventing the removal of these materials from their farmlands. Our presentations and attitudes should reflect the fact that we recognize the commonality of our interests.

6.   CONCLUSION

The Great Barrier Reef World Heritage Area, including the Marine and National Parks which constitute its major management regimes, is a working example of Integrated Coastal Management and the practical application of the principles defined in the World Conservation Strategy. It can be seen as a model for development of the kind described in the report of the World Commission on Environment and Development - "Our Common Future" (World Commission on Environment and Development, 1987).

7.   REFERENCES

Australian Travel Industry Association (ATIA), 1984. Data review of reef related tourism 1946-1980. Townsville, Great Barrier Reef Marine Park Authority Publication, 120 p.

Commonwealth of Australia, 1980. Offshore constitutional settlement. A milestone in cooperative federalism. Canberra, Australian Government Publishing Service, 18 p.

Europa World Year Book, 1995. Australia. London, Europa Publications, Vol.1, pp 407-36

Darwin, C., 1906. The origin of species. London, J. Murray, 703 p.

Great Barrier Reef Marine Park Authority (GBRMPA), 1994. Keeping it great. The Great Barrier Reef. A 25-year Strategic Plan for the Great Barrier Reef World Heritage Area, 1994-2019. Townsville, GBRMPA, 64 p.

Moss, A.J., G.E. Rayment, N. Reilly and N.K. Best, 1992. A preliminary assessment of sediment and nutrient exports from Queensland coastal catchments. Report of the Queensland Department of Environment and Heritage, the Department of Primary Industries. Queensl. Dept.Environ.Herit.Tech.Rep., (5):33p

Resources Assessment Commission, 1992. Coastal zone inquiry. Background paper. Canberra, Australian Government Publishing Service, 69 p.

World Commission on Environment and Development, 1987. Our common future. Oxford, Oxford University Press, 383 p.

Zagorin, A., 1995. Real Wealth of Nations: a thought provoking report from the World Bank ranks nations by a ?greener? set of standards. Time, 2 October 1995, p. 36

Zann, L.P. (comp.), 1995. Our Sea, our future. Major findings of the state of the marine Environment Report for Australia. Townsville, Australia. GBRMPA for the Department of the Environment, Sport and Territories, Ocean Rescue 2000 Programme, 112 p.

Annex 3

CASE STUDY 3 - ECUADOR'S COASTAL RESOURCES MANAGEMENT PROGRAMME

by Stephen B. Olsen[1]

Coastal Resources Center, University of Rhode Island, Bay Campus, Narragansett RI 02882, USA

1. THE CONTEXT

1.1 Some Characteristics of Ecuador and its Coastal Region

Ecuador's coastal region, when defined to include the four provinces that roughly coincide with the coastal plain, is the country's fastest growing region. With a shoreline of 1,256 km and a population of 4.7 million in 1990 growing at about two percent per year, the population doubles about every 35 years. In 1990, fifty percent of the population was classified as under-employed, and the proportion living in poverty was gradually increasing. Approximately 15,000 artisanal fishermen depend directly on estuarine and coastal fin fisheries and a similar number harvest shrimp post-larvae to stock the 146,000 ha of shrimp farms (Epler and Olsen, 1993). While Ecuador's most valuable export is petroleum produced in the Amazon basin, all other important industries are coastal and include bananas, farmed shrimp, cacao and some coffee. The coastal region has been transformed in the past forty years first by rapid population growth and urbanization, the removal of all but small remnants of once extensive coastal forests, and most recently by a boom in shrimp farming that has radically changed the estuaries and associated mangrove wetlands in all but the northernmost coast. In 1992, 116,315 t of farm-grown shrimp were produced.

With a small manufacturing industry, Ecuador's economy is dependent upon the export of unprocessed raw materials. While oil production in the interior promises to continue to provide the largest source of income, other exports and the livelihoods of the majority of the population are dependent upon the sustained yields of agriculture, fisheries, mariculture, and timber. The wealth produced by these activities is concentrated within the relatively small middle class and the much smaller elite.

1.2 The Pre-existing Coastal Governance Structure and Its Effectiveness in Implementing Resource Management Policies

Before Ecuador's coastal management programme got underway in 1986, the nation's reliance upon its endowment of natural resources had been recognized by a succession of governments and numerous laws. Implementing mechanisms were in place to control or halt activities that degraded important ecosystem qualities or resulted in over-exploitation. However, the implementation of research, planning and regulation had, in almost all cases, a negligible impact on the increasingly urgent problems posed by inappropriate shorefront development, declining water quality, over-fishing, deforestation and soil erosion. The failure to connect scientific knowledge of how coastal ecosystems function and can be sustained to public policy in order to halt or slow destructive behaviour, demanded reassessing the process by which such public policy is formulated and implemented. In Ecuador, the quality of scientific information relevant to such issues did not appear to be the factor that limited effective coastal management.

In Ecuador, governmental authority is concentrated within central government. Responsibility for policy formulation, development, planning, regulation, research and extension on activities and resources in the coastal region is fragmented among five ministries and several

---

[1] Co-Director of Ecuador's Coastal Resources Management Programme, 1986-1993

agencies within each ministry. Planning and policy formulation are coordinated by an agency (CONADE) chaired by the vice-president, but the influence of this agency varies from one administration to another. Provinces are administered by governors appointed by the President who play some role in coordinating activities of the many governmental agencies. Municipal government is very often weak and usually has little control over the development process. Along the mainland coast, a branch of the navy, DIGMER, regulates, through a permit system administered by port captaincies, some forms of shorefront construction, dredging and related activities.

2. ECUADOR'S COASTAL RESOURCES MANAGEMENT PROGRAMME

2.1 The Goal and Objectives of the Programme

In 1986, the Governments of Ecuador and the United States signed an agreement, the goal of which was to establish a national coastal management programme for the mainland coast. From 1986 to 1993, the project was administered by the University of Rhode Island's Coastal Resources Center (CRC) as one of three pilot projects designed to test how the concepts and techniques of coastal management could be applied and advanced in developing tropical nations. While the original agreement set objectives modeled on the regulatory approach that characterizes state coastal management programmes in the U.S. (e.g., formalizing an impact assessment process for major proposals, zoning critical areas, developing shorefront construction standards), the project design called for an incremental, learning-based approach that featured annual assessments of success and failure and reassessments of programme strategies by the full Ecuadorian-American team. Project funds were allocated through annual work plans based upon such assessments and approved by the project's American and Ecuadorian co-directors. CRC's partner in the Ecuadorian government was the Office of Natural Resources Management (until 1989) and, subsequently, the Directorate of Public Administration in the Office of the President of the Republic. These first eight years of the programme were funded by USAID and the Government of Ecuador. Beginning in 1996, a four-year implementation phase will be supported by a loan from the Inter-American Development Bank. Thus, the first generation of Ecuador's Programa de Manejo de Recursos Costeros (PMRC) will span a fourteen-year period *(see Table I)*.

After an initial year of adjustment, the objectives of Ecuador's first generation coastal management programme were revised to focus upon the need to enhance the institutional and societal context for improved resource management. Specifically, the objectives of the programme became to:

* create and mobilize constituencies for improved coastal management both at the community level and within central government;

* formally establis governance structures and processes at the community level and within the central government by which improved management can occur;

* build indigenous capacity to deliver improved management;

* experiment with resource management techniques at a pilot scale in five special area management zones to discover and demonstrate promising approaches to priority coastal management problems.

Table I
Stages and Events in the ICM Process
in the Ecuador Coastal Resources Management Programme

Stages in Policy Process	Year	Key Events
STAGE 1 Issue identification and assessment	1986	USAID-GOE-URI coastal resources management project agreement; Shrimp mariculture assessment and national symposium
	1987	Compilation of available information on the condition and use of coastal resources: • Overview of the coastal region • Province profiles and workshops
STAGE 2 Programme Preparation	1988	Public review of information, definition of objectives for a national CRM programme; Institutional structure and programme priorities proposed
STAGE 3 (a) Formal Adoption	1988-89	Manifesto of support signed by local leaders; Executive Decree 375 formally creates the Ecuador Coastal Resources Management Programme: • Technical Secretariat • Special area management zone (ZEM) offices • Ranger Corps units
(b) Detailed Planning and Securing Funding	1990-91	Preparation of ZEM plans; Practical exercises in integrated management; Initiation of the Ranger Corps
	1992	Approval of ZEM plans; Executive Decree 3399 to restructure and decentralize the PMRC; Inter-American Development Bank (IDB) loan proposal
	1993	Review and approval of the loan; documentation of PMRC experience
STAGE 4 Full-Scale Implementation	1996 -2000	Implementation through the IDB loan programme
STAGE 5 Evaluation	1999 -2000	Overall programme evaluation of progress in the PMRC's work, including problems and lessons learned, and definition of issues and scope for a second-generation coastal programme.

Formal enactment of the Programa de Manejo de Recursos Costeros (PMRC) occurred in 1989 when President Rodrigo Borja signed an Executive Decree that created the National Commission for Coastal Resources Management comprised of representatives of the five ministries directly involved in coastal management. The Commission is charged to formulate national policy on coastal management issues and resolve coordination issues among governmental agencies. A secretariat staffs the Commission and is responsible for administering five Zonas Especiales de Manejo, or ZEMs (Special Area Management Zones), selected as microcosms for the combinations of resource management issues typical to urban and rural areas along the mainland coast. Rather than attempting to operate along the entire coast, the PMRC is focusing its efforts in these five ZEMs during this first generation effort. The decree also created the Coastal Ranger Corps that draws together, under the leadership of Navy Port Captains, the local level regulatory and enforcement functions of several ministries. Each ZEM operates through a resident coordinator and the local planning and conflict resolution process is overseen by ZEM committees that draw together representatives of both user groups and local government. A non-goernmental organization (NGO), the Fundación Pedro Vicente Maldonado, has played a key role in the programme, overseeing the initial issue analysis and prioritization process at the provincial level and subsequently organizing ZEM level activities. This NGO also administers the public education programme. The structure of the programme is shown in Figure 1.

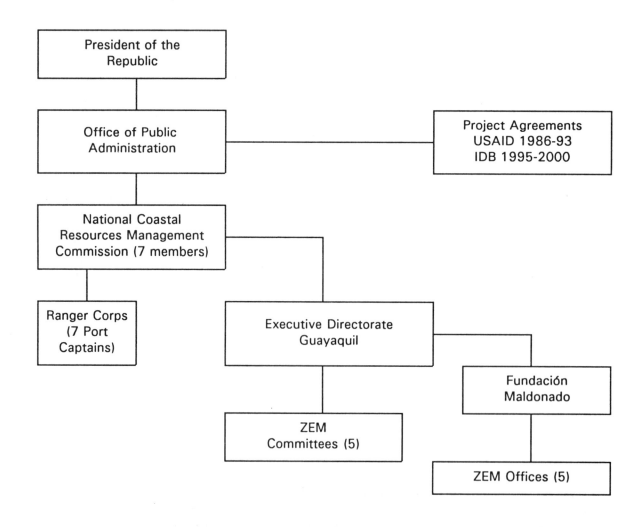

Figure 1. Structure of the PMRC under Executive Decree 3399 of 1992

2.2   Major Strategies of the Programme

Ecuador's first generation coastal management programme has addressed five issues:

- governance structure and process,
- mangrove destruction,
- mariculture expansion,
- shorefront development, and
- environmental sanitation.

These issues have all been addressed following a two-track strategy. This calls for simultaneously building constituencies both within central government and within the ZEMs to manage the coastal development process more effectively. The power of this approach lies in creating a dialogue that links the two tracks and promotes a sense of shared purpose at both levels. A strong and well informed first track is essential to ensure that greater responsibility and initiative at the local level is not perceived as a threat to the power and prerogatives of central government. By placing the programme in the Office of the President in 1989, the first generation initiative was able to command the attention and cooperation of the various ministries and gain attention at the local level. The programme draws upon the regulatory authority vested in the agencies within the collaborating ministries. Thus, in this initial generation, the programme has specifically avoided a formalized redistribution of agency authority and has focused upon improved coordination and efficiency within the existing institutional framework.

Table II summarizes the major strategies that emerged as most useful during the initial years of the programme as it worked to set the stage for more effective coastal management. A full description and assessment of stages by which the programme has evolved and the strategies adopted to address the major management issues are set forth in Robadue (1995) and summarized in Table II.

2.3   The Geographic Boundaries of the Programme

The Stage-One issue assessment considered a broadly defined coastal region that includes the four provinces that roughly coincide with the coastal plain and cordillera between the base of the Andes and the Pacific Ocean. During the second phase of planning, attention was directed almost exclusively on the five ZEMs. These have shorelines ranging from 25 to 75 km in length. Inland ZEM boundaries were defined in only general terms since the "findings of fact" policies and actions detailed in each ZEM Plan are framed by issue and not by rigid geographic boundaries. Where feasible, each plan analyzes issues and proposes actions for the watersheds that include both the coastline and portions of the associated cordillera.

3.   THE CONTRIBUTIONS OF SCIENCE AND SCIENTISTS TO THE ISSUES ADDRESSED BY THE PROGRAMME

Ecuador provides an example of a nation where societal and institutional values and structures range from weakly positive to strongly negative for improved coastal management. During this first generation, the PMRC has therefore concentrated its efforts on improving the social and institutional context for resource management and experimenting with small-scale management initiatives to discover which techniques build constituencies and show the greatest promise.

3.1   Governance Process and Structure

This issue has been the major focus of the PMRC. As suggested by Figure 2, improvements in governance cannot occur unless the structures and procedures by which such governance occurs are strong enough, and adequately motivated, to produce implementation.

Table II
Strategies to Address Priority Expressions of Coastal Degradation

---

MANGROVE DESTRUCTION

Strategy 1: Increase public awareness of the benefits produced by mangrove ecosystems and the alarming trends of losses in their abundance and condition

Strategy 2: Develop and test mangrove management techniques that promote community-level stewardship and sustained use

Strategy 3: Improve awareness and enforcement of mangrove laws and regulations

Strategy 4: Work with the national agencies that are responsible for mangrove management to prepare a proposal for a new approach that emphasizes planning and sustained use at the community level

Strategy 5: Apply international experience to foster monitoring and research in support of management

MARICULTURE EXPANSION

Strategy 1: Prepare and promote a vision for a sustainable mariculture industry for Ecuador

Strategy 2: Bring international experience to bear in addressing mariculture issues

Strategy 3: Take actions at the local level to protect the environmental base of the mariculture industry

Strategy 4: Diversify the flow of benefits and species cultured

SHOREFRONT DEVELOPMENT

Strategy 1: Map all coastal features, analyze problems and opportunities for use of the shore, and prepare recommendations on good development practices

Strategy 2: Focus efforts to prepare and implement shore use plans and zoning in the five ZEMs

Strategy 3: Examine the economic and marketing potential of recreation and tourism development, especially in terms of its link to good environmental quality

ENVIRONMENTAL SANITATION

Strategy 1: Design and implement a water quality sampling programme focused on issues relating to shrimp mariculture, assess the strengths and capacity of in-country water quality laboratories and combine the results of these efforts into the design of an integrated programme to generate baseline data and conduct continuous water quality monitoring

Strategy 2: Guide and promote public investments and private collaboration in pollution control in the ZEMs

The PMRC has proceeded on the assumption that its first priority, in terms of science and scientists, is to create an institutional and societal context for more effective coastal management that can receive and respond to new information and ideas on how coastal ecosystems function and respond to human actions. Until a more favourable context has been created, the PMRC has reasoned, investing its own limited financial resources in original research in the natural sciences will be of limited usefulness. The emphasis has instead been upon making the fullest possible use of scientific concepts and applicable information on the processes that govern ecosystem function and the likely consequences of the human activities addressed by the PMRC.

INTERMEDIATE OUTCOMES		FINAL OUTCOMES	
FIRST ORDER	SECOND ORDER	THIRD ORDER	FOURTH ORDER
Programme preparation; Constituencies and formalized structures for ICM	Initial implementation; Correction and mitigation of selected behaviours	Improvements in condition/use of target resources or attributes	Improved environmental quality and quality of life

Figure 2. Ordering Coastal Governance Outcomes (Adapted from: United States, 1994)

This strategy was reinforced by the experience of assembling a team of internationally respected social and natural scientists in the first year of the project to analyze the factors affecting the sustainability of the shrimp mariculture industry. The priority factors identified, and the predictions made by this interdisciplinary group have proved to be remarkably accurate over the subsequent decade. As foreseen by their analysis, mounting production problems caused by declining water quality; increasing scarcity of wild-caught seed shrimp and the egg-bearing female shrimp required by hatcheries; and shifts in world shrimp production and markets have all combined to produce a series of crises that make it all too likely that this industry will repeat the boom-bust tradition of natural resource exploitation in Ecuador. Thus, while the scientific analysis was accurate and timely, neither the government nor the shrimp industry was willing or able to respond to the recommendations made by this group of social and natural scientists. This is the fate of innumerable relevant, timely and technically-sound studies and plans in the developing world.

During Stages One and Two, the PMRC focused its analysis upon:

- the synthesis of existing information on priority coastal management issues as derived from secondary sources, interviews and public debate;

- the existing legal and institutional framework as it applies to those priority issues.

The Ecuadorians and Americans that worked together on these topics were social scientists, legal scholars and resource managers. The questions addressed by these two groups of analyses were:

ISSUE ANALYSIS

(1)   What are the historical trends in the condition and use of important coastal resources?

(2)   What are the social, economic and environmental forces at work that drive these trends?

(3)   What are the societal implications of such trends?

(4)   How good are the existing data as a basis for framing public policy?

(5)   What are the prevailing public perceptions of the importance and implications of such trends—can differences in such perceptions be linked to different user groups or sectors of society?

LEGAL AND INSTITUTIONAL ANALYSIS

(1)   What are the legal mandates, responsibilities and formal policies of the governmental agencies with responsibilities for coastal management? How do these overlap, contradict or support one another for each of the priority coastal management issues?

(2)   What is the capacity of the most relevant institutions to improve their performance?

(3)   What are the perceptions of high level officials in each agency as to the nature of the problem and potentially productive courses of action to improve how coastal resources are managed?

3.2   Mangrove Destruction

When the coastal management partnership got underway in 1986, there was already widespread concern that the extensive mangrove wetlands around Ecuador's many estuaries were being rapidly destroyed by shrimp farms. Governmental officials and many participants in the industry, recognized the importance of mangroves as a juvenile fisheries habitat, as a storm buffer, and as a source of livelihood to many thousands of the poorest coastal dwellers. The major questions addressed by the PMRC which required scientific knowledge were:

(1)   What is the magnitude of the mangrove wetlands, how are they distributed and how fast are they being destroyed? This information was provided by the Remote Sensing Unit of the Armed Forces Mapping Office (CLIRSEN), which received training in photo interpretation and ground proofing by an American university and the French Foreign Assistance Programme. CLIRSEN produced detailed maps that classified mangroves into various height categories and quantified losses by estuary between successive surveys.

(2)   What are the specific values and services provided by mangroves? The technical complexity of answering this question in quantifiable terms specific to Ecuador was judged to be beyond the capability of the PMRC. Here the PMRC's strategy was to make available in Ecuador international research on this topic. When research on aspects of this topic was funded in Ecuador by another programme, the PMRC provided office space, transport, etc., to visiting scientists, and encouraged mentoring relationships between foreign scientists and local scientists to promote local interest and strengthen local capabilities.

The government's approach to accelerating mangrove destruction during the eighties was to adopt ever more stringent and more unenforceable regulations designed to prohibit or severely limit human activities of all kinds in mangrove wetlands. The PMRC has pioneered a different strategy that is now gaining considerable support among both government agencies and the public. The strategy calls for developing mangrove management techniques that promote a diversity of sustainable mangrove activities organized and administered at the community level.

These activities include the construction of simple walkways into mangrove wetlands to permit educational and ecotourism tours, sustained use of designated mangrove areas for charcoal production and wood cutting, and zoning schemes that allocate specific mangrove areas for specific fishing activities. While these activities have been supported by technical advice from international advisors on replanting techniques and sustainable harvesting regimes, the leadership of these experiments has been provided by a local forester. User agreements negotiated between traditional fishermen, woodcutters and shrimp farm owners have proved to be a powerful tool in resolving conflicts that otherwise frequently lead to violence. PMRC-sponsored user agreements are endorsed and enforced by the local coastal Ranger Corps. These activities, complemented by sustained public education and coordinated patrols by the user groups, have led to a sharp decline or cessation of mangrove destruction within the five ZEMs. In some instances, illegal cutting and dredging by shrimp farmers has led to negotiations that have resulted in the transgressor agreeing to replant a larger area than the one destroyed. The National Commission has endorsed such initiatives and encouraged the interagency collaboration they require.

3.3   Shorefront Development

Bt 1986, the shrimp farming boom in Ecuador had resulted in the engineering of all the country's estuaries with the exception of the largely inaccessible northern coast of Esmeraldas province. This development has resulted in channelizing water flow, often radically altering water circulation and exchange and making the coast more vulnerable to damage during the storms associated with severe El Niño events. The rapid growth of cities has produced large squatter settlements built out over intertidal areas many of which subsequently are filled. Along the open coast, roads, second-home developments and hotels have been built close to or directly under unstable cliffs, near inlets and within the swash zone of major storms. In this case, the PMRC posed the following technical questions:

(1)   What are the geomorphological processes at work on the Ecuadorian coast that should be considered as the development process proceeds?

(2)   Where has existing development posed significant problems of erosion and accretion, or present future hazards; where has shorefront development caused conflicts among user groups of significantly reduced public access to the shore?

(3)   What are the likely causes of existing coastal erosion and accretion problems and how should they be addressed?

(4)   What are the specific priorities for action and specific guidelines for future development for each segment of the coast?

(5)   What specific sites and topics require further research and/or monitoring?

These questions were addressed by a two-person team comprised of an American coastal geomorphologist with many years of experience in analyzing the impacts of human development on shoreline processes and an Ecuadorian geologist. In this case, the PMRC sponsored a survey of the entire coastline that produced the first detailed atlas of the immediate shoreline. The atlas identified the physical processes at work and problem areas caused by existing development, and made specific recommendations for each segment of the coast to guide future development.

3.4   Environmental Sanitation

Public health is at the top of the list of concerns in all coastal communities and within the ZEMs. Diseases associated with poor water quality are the major source of mortality. Furthermore, shrimp farmers are increasingly aware that the reduced production of many shrimp farms appears to be associated with changes in water quality. Building upon the recommendations of the 1986 symposium on a sustainable mariculture industry, the PMRC's Water Quality Working Group has addressed the following questions:

(1)   What is the current status of water quality in Ecuadorian estuaries and along the open coast as documented by reliable survey techniques?

(2)   Where are the most significant water quality problems and what specific pollutants are of concern?

(3)   What are the likely sources of such pollution?

(4)   What are the priorities for immediate action?

(5)   What are the priorities for further research and monitoring?

The compilation of existing water quality information demonstrated that there were major discrepancies in data for the same variables in the same area and major questions on the quality of such data. The first priority was to improve the reliability and overall quality of indigenous laboratories present in several universities and a number of governmental agencies. A Water Quality Working Group was formed and provided with a foreign advisor. The Water Quality Working Group began by undertaking a series of intercalibration exercises with collaborating laboratories in the United States. This identified weaknesses in technique and priority needs for upgrading equipment. These exercises also led to a natural sorting of the capabilities and expertise for analyzing different variables amongst the different laboratories. Once greater confidence had been achieved in the quality of the data, potential problem areas were sampled. This carefully-targeted sampling programme revealed geographic areas of concern and provided an incomplete but better quality baseline against which to measure future trends. The work of the Water Quality Working Group has documented pervasive problems in eutrophication and low oxygen conditions in areas where shrimp ponds are dense and water flow reduced. Mercury pollution that is advancing towards estuaries in streams from hills where primitive forms of unregulated gold mining are underway is another important issue addressed by these surveys.

4.   ENHANCING THE SUSTAINED CONTRIBUTIONS OF THE SCIENCES

4.1   Building Indigenous Capacity

In the initial stages of the PMRC, there was little appreciation within the community of local natural scientists for what information would be most relevant to a coastal management initiative. The PMRC received numerous unsolicited proposals for studies of offshore hydrography, inventories of various categories of biota and the estimation of, for example, the potential for a new export crab fishery. Many researchers were mystified, if not angered, when such proposals were turned down. In order to build capacity for scientific work that would be of direct usefulness to a better understanding of priority issues, the PMRC adopted four strategies:

Strategy 1:   Form inter-institutional working groups on important coastal management issues

Strategy 2:   Foster long-term mentoring relationships with scientists of international stature

Strategy 3:   Focus PMRC-funded research on questions of direct relevance to the coastal management programme

Strategy 4:   Improve the quality and reliability of Ecuadorian research on policy-relevant topics

A sustained effort to implement these four strategies has built interest and capability in policy-relevant science in coastal Ecuador and created a more favourable context for sustained ICM efforts.

4.2   Creating a Demand for High Quality, Policy-Relevant Science

By 1993, the institutional structures, the constituencies, and specific ZEM plans were all in place. The last necessary ingredient was the funding with which to move forward. That year the Ecuadorian government identified the continued funding of the PMRC and the implementation of the ZEM plans as a top priority to the Inter-American Development Bank (IDB). The Bank's response was positive and subsequently designed a loan programme to support the implementation phase of the first generation programme.

Now that a more favourable institutional and social context is in place that has the ability to respond to new scientific information, a greater emphasis upon PMRC-sponsored scientific research is in order. The ZEM plans formally adopted at the community level, and subsequently endorsed by central government, contain broad statements of policy on the five priority issues and sets of "first-step" actions. Implementation of these policies and actions are judged to be within the capabilities of the newly created coastal management institutions with some minimal outside technical assistance and modest funding. The existing ZEM plans are not based upon a systematic analysis of the ecosystems concerned nor do they attempt to identify, for example, what measures might restore water circulation in a given estuary or how much mangrove should be restored and where. The existing ZEM plans take only the first step, by calling for a halt to further shrimp pond construction and testing the feasibility of community-implemented mangrove replanting at a pilot scale.

As the implementation phase of the first generation PMRC programme gets underway, the PMRC will begin to lay the groundwork for "second generation" ZEM plans based on more sophisticated analyses. This will require considerable scientific research that will provide, among other things, a basis for setting specific quantitative objectives for the development and conservation issues addressed. Thus, approximately ten percent of the IDB budget is allocated to three years of research that will provide the foundation for a comprehensive plan for the Rio Chone estuary and its immediate watershed. The hope is that by conducting carefully defined research on the processes that govern the qualities of this relatively small estuary, the indigenous scientific expertise, complemented by selected outside advisors, can learn to function as an effective interdisciplinary partnership. If this can be accomplished for the Rio Chone, then the far more complex and larger estuaries to the north and south, most importantly the Guayas itself, could be tackled.

While the science sponsored by the PMRC loan programme will be focused upon the Rio Chone, small supplementary investments will be funded by the loan to further document the by-catch of the post-larvae fishery, and to assess the potential impacts of the fishery for egg-bearing female shrimp on shrimp stocks.

A third priority in the IDB-sponsored phase of the PMRC programme is to document baseline conditions and monitor change both in the governance process itself and the condition and use of those resources and environmental qualities that the programme is attempting to conserve. Unfortunately, such baselines were not developed at Stage One of the programme in 1985-1986. A mix of social and natural sciences will be required to select the most useful questions, identify indicators and design methodologies.

5.   MAJOR ACCOMPLISHMENTS TO DATE

The USAID-sponsored phase of the PMRC was designed as a policy and planning initiative that did not provide funding for implementation. A major goal was reached with the formal enactment of the program through Executive Decree 375 in 1989. Constituencies that actively support the program's new governance structures and process have been created and sustained at both the community level and within central government through three presidential administrations. The biggest advances in creating new governance structures are with the ZEMs and the Ranger Corps. The "practical exercises" strategy has provided for the transition from planning to implementation at a pilot scale for all issues addressed. All ZEMs contain examples of implementation ranging from functioning protected areas to community built sanitation projects. Both the governance procedures and ZEM level implementation initiatives have been sustained over a two-year period when external funding was interrupted.

6.   SOME LESSONS LEARNED

(a)   Ecuador's PMRC demonstrates the importance of matching new investments in social and natural sciences to the willingness and ability of a society to respond to new information and new ideas. A central lesson is to recognize there are major differences among nations in their ability to respond effectively to scientific knowledge and the advice of scientists. Scientific knowledge and information are often not the factors limiting progress towards more effective coastal management. This is particularly true for developing nations where the implementation of adopted public policy on natural resource management is weak.

(b)   In first generation programmes, an emphasis upon existing data, rather than commissioning new studies, brings many benefits:

- it focuses attention on analysis rather than data gathering
- it encourages seeking out and involving indigenous expertise
- it encourages consideration of trends rather than "snapshots"
- it is less costly than commissioning primary research
- it encourages careful targeting of research on policy-relevant questions

(c)   Learning is much enhanced if baselines for public perceptions, the governance process and the quality and use of natural resources are documented at the start of the initiative and then monitored. Explicit statements on the hypotheses that underlay the design of the project and its priorities also hasten the learning process.

(d)   The most useful roles for the non-indigenous scientific "expert" is to motivate, verify and occasionally provide technical assistance to their local counterparts. Long-term mentoring relationships are particularly beneficial.

(e)   External experience, perspectives and conceptual frameworks can be extremely helpful once a social and institutional context exists wherein new ideas can be received and acted upon. Systems analysis and systems thinking is particularly powerful in breaking down traditional sector-by-sector, discipline-by-discipline analysis and actions.

(f)   Scientific research and analysis benefits from a strong issue-driven approach that focuses attention on selected policy-relevant questions.

(g)   Inter-institutional and interdisciplinary, issue-specific working groups can be a powerful means for promoting systems approaches to the application of science to problem solving.

7.   REFERENCES

Epler, B., and S.B. Olsen, 1993. A profile of Ecuador's coastal region. Coastal Resources Center, University of Rhode Island. CRC Tech.Rep., (2047):139p

Robadue, D. (Ed), 1995. Eight years in Ecuador: the road to integrated coastal management. Coastal Resources Center, University of Rhode Island; U.S. Agency for International Development; Global Environment Center. CRC Tech.Rep., (2088):319 p.

United States, Environmental Protection Agency, 1994. Measuring progress of estuary programs. Washington DC, U.S.EPA, Office of Water, Doc. 842-B-94-008, 267 p.

Annex 4

CASE STUDY 4 - COASTAL MANAGEMENT IN BOLINAO TOWN AND THE LINGAYEN GULF, THE PHILIPPINES

by Edgardo D. Gomez and Liana McManus
Marine Science Institute, University of the Philippines, Diliman, Quezon City, Philippines

1.   BRIEF DESCRIPTION OF THE CONTEXT FOR THE PROGRAMME

1.1   Salient Characteristics of the Philippines and Trends in the Condition and Use of Coastal Ecosystems

The Philippines is an archipelago of some 7,000 islands in the western portion of the Pacific Ocean just north of the Equator. It has a coastline of approximately 36,300 km. The Philippines forms the eastern boundary of the South China Sea, setting this off from the broader Pacific Ocean. It is entirely tropical in climate, experiencing a reversing monsoon regime with pronounced wet and dry seasons. The region boasts the most diverse coral reef systems in the world, being the centre of their development and diversity. Coastal waters support a substantial fishery, both for local consumption and for commercial purposes. Unfortunately, the demand for living resources is increasing, resulting in more stress on the coastal environment.

There is a great dependence on the marine environment and marine resources. The concentration of populations in the coastal zone often leads to land- and sea-use conflicts. Many natural habitats have been degraded or lost. Reefs are being degraded by blast fishing, sedimentation and other forms of pollution, and extractive activities. Mangrove wetlands have been severely exploited. Of the original half a million hectares, less than a quarter remain in their original condition. While it is claimed that twice this area is still mangrove covered, the growth is secondary and of varying quality. About 2,000 km^2 of mangroves have been converted into ponds for fish and shrimp culture. This situation is not atypical for the region, although large tracts of forested wetlands occur in Indonesia.

The Lingayen Gulf, a semicircular embayment with an area of about 2,100 km^2 on the northwestern coast of the main island of Luzon, has been described by McManus and Chua (1990). About two-thirds of the 160 km coastline is sandy (east and south) while the western shoreline is fringed by a major coral reef system opening to the South China Sea. Seaward of these reefs, the central gulf has an average depth of 46 m, a sandy-muddy bottom, and supports a trawl fishery. The reefs are heavily fished, both for finfish and other marine products. The surrounding land area is primarily agricultural, both on the eastern shore in the province of La Union and on the southern and western shores which are part of the province of Pangasinan. Coastal tourism is important in the gulf which contains one of the oldest national parks in the country, the Hundred Islands of the Alaminos municipality.

At the northwestern end of the gulf is the municipality of Bolinao, one of the eighteen municipalities that border upon the Gulf. This municipality of 50,000 inhabitants is an agricultural centre with a fishing industry that provides livelihoods to about a third of the population. The condition of the marine resources of the town is described by McManus et al. (1992). The reef fishery is overexploited and management measures are urgently needed to restore these resources. Numerous articles have resulted from the research studies conducted at the University of the Philippines Bolinao Marine Sciences Institute that document and analyze the marine resources in the vicinity and examine management issues.

1.2   Wealth of Society and its Dependence on Coastal Ecosystems for Livelihoods

The Philippines is at the initial stages of industrialization. Agriculture employs nearly half of the population. A large proportion of coastal dwellers depend directly upon the local coastal resources for their food supply and livelihoods. There are more than 600,000 municipal fisher families in the Philippines and they are one of the poorer sectors of society. A high population growth rate aggravates their condition and their future prospects.

The municipality of Bolinao contains 30 villages or barangays, 14 of which have coastlines. These coastal barangays contain 60 percent of the town's population. The town is typical of rural Philippines and the economic condition of the people needs substantial improvement. There is no starvation and there are no serious health problems but 88 percent of the population live in poverty. About half of the work force is dependent on farming while about one third is involved in fisheries. Since many of the latter are not land owners, they depend heavily on marine resources for their livelihood.

1.3   Existing Governance Structure and its Prior Effectiveness in Successfully Implementing Natural Resource Strategies

The Philippines enjoys a democratic form of government that was patterned on the United States model. A traditionally strong central authority divides the country into thirteen regions, each of which is made up of several provinces headed by a governor. Philippine politics has been characterized as revolving more around key personalities rather than around fixed institutional structures. In the decision making process, there is a balance between the Rule of Law and the Rule of Man that strongly influences how natural resource management policies are implemented and how important decisions are made. In recent years, there has been a slow devolution of power to local governments. In 1991, the Local Government Code was enacted giving municipalities primacy in the governance of their local affairs, including the management of natural resources. The devolution of power has been a slow process because of the lack of clear implementing guidelines and the inconsistencies with a legal system that remains oriented to a centralized government. Hence, there is some instability in the governance system while the municipalities begin to understand and exercise greater authority.

2.   DESCRIPTION OF INITIAL ICM PROGRAMMES

In the mid-eighties the regional **ASEAN/US Coastal Resources Management Project** (CRMP) sponsored pilot programmes in coastal management in each of the six participating countries. The Lingayen Gulf was chosen as the Philippine project site in part because of a significant body of knowledge for the gulf and the presence of the Bolinao Marine Laboratory of the Marine Science Institute. The project ended in 1992 with the main output being the Lingayen Gulf Coastal Area Management Plan (the Gulf Plan) as endorsed by the National Economic and Development Authority, the government's national planning body (NEDA, 1992). The implementation stage of the plan was delayed until 1994, when the president of the republic, Fidel V. Ramos, issued Executive Order No. 171 creating **the Lingayen Gulf Coastal Area Management Commission** (the Gulf Commission) and charged it to implement the plan.

The preparation of the management plan prompted interest within the Institute to address coastal management issues and to collaborate with the University's College of Social Work and Community Development, and a non-governmental organization (NGO), the Haribon Foundation, in exploring integrated and community-based approaches to coastal resource management. This collaboration in turn gave rise to a companion project—the **Bolinao Community-Based Coastal**

Resources Management (CB-CRM) Project that was funded in late 1993 by the International Development Research Center of Canada (IDRC).

During the formulation of the CB-CRM Project proposal, an unexpected development greatly increased the need for an effective planning and management process. Local politicians and foreign investors were and still are planning a large cement factory, apparently without any consideration for the ongoing coastal management studies and initiatives. Suddenly, Bolinao town and the Lingayen Gulf became a major environmental issue for the government, pitting some local interests against moneyed developers and their political backers. The cement plant proposal made the transfer of sound technical information a priority for the CB-CRM project so that all parties could better assess their development options.

2.1   Goals of the Programmes

The primary goal of the ASEAN-US Coastal Resources Management Project was "to strengthen the capability of ASEAN countries to develop their renewable coastal resources on a sustainable basis. .. [and] to help ensure the long-term productivity of coastal fisheries and aquaculture, mariculture, forestry and other forms of primary resource dependent development" (ASEAN-US Project Memorandum of Understanding, 1986).

Four institutions were identified in 1986 to implement the Philippine component of the ASEAN-US Project: the University of the Philippines Marine Science Institute (UP MSI), the UP Visayas College of Fisheries (UPV CF), the UP College of Social Work and Community Development (UP CSWCD), and the Bureau of Fisheries and Aquatic Resources (BFAR). All four institutions had capabilities to undertake an assessment of resources and habitats in the gulf. They did not, however, have experience in formulating an integrated coastal management plan. In the fourth year of the project, the National Economic Development Plan (NEDA)-Region I was designated as the lead agency for the formulation of the gulf-wide management plan.

The goal of the resulting Lingayen Gulf Coastal Area Management Plan (Gulf Plan) is "...social and inter-generational equity in the use of coastal resources, poverty alleviation, enabling legal arrangements, and collective advocacy..." through the mechanisms and strategies of integrated coastal management. Specific plan objectives are directed at the rationalization of land and water use through zonation schemes, the rehabilitation of degraded coastal habitats, livelihood development, law enforcement and the provision of infrastructure and support service mechanisms for the management of the gulf.

The Gulf Plan addresses a large area that includes 18 coastal municipalities, one city and five non-coastal towns with estuarine fishponds. The outermost limit of the study area is the 200 m isobath and the terrestrial boundary is 1 km from the 160 km long shoreline. The total area considered by the plan is 2100 km^2 of the gulf.

Unlike the Lingayen Gulf which has a management plan that is carried out by a Commission, Bolinao as yet has no written management plan. As a municipality that is heavily dependent on its marine resources, Bolinao has responded to the declines in its resource base through resource-specific interventions by municipal legislation, which is, however, influenced by the need to raise revenues through taxation and fees for harvest and transport licenses, rather than by the need for resource protection. The goal of the CB-CRMP is to provide the municipality and its villages with the tools of coastal management.

Five villages were chosen for the initial phase of the CB-CRMP programme. Although the programme concentrates on coastal villages, management plans all have to be made within the

.context of a town plan since this is the smallest scale at which resource management is legally mandated. Village resolutions can be drafted and passed but will be promulgated only when approved by a Municipal Council.

2.2 Major Objectives of the Bolinao Community-Based Coastal Resources Management Project

For the CB-CRMP the objectives are:

(a) Community organization, environmental education and institutionalization focused on empowering local communities to function effectively as stewards of their coastal resources.

(b) Evaluation of resource use and management options through participatory research including: marine protected areas, habitat restoration, aquaculture technology, land-based production systems, and coastal development plans.

(c) Livelihood development to increase levels of food production and generate income.

(d) Networking and advocacy to establish linkages with other similarly motivated groups and agencies to institutionalize goals and advocate change in policies and laws within higher levels of government.

3. MAJOR ISSUES ADDRESSED BY THE PROGRAMMES

3.1 Overview

The priority issues addressed by both the Gulf plan and the CB-CRMP are almost identical and may be listed as follows:

LINGAYEN	BOLINAO
Overfishing	Overharvesting of reef resources
Destruction of critical habitats	Degraded coral reefs
Pollution	Increasing sewage pollution
Unsustainable aquaculture practices	
Coastal erosion	
Lack of alternative livelihood	Lack of alternative livelihood
Weak institutional arrangements	
Lack of political will	Low level of environmental awareness

3.2 The Significance of the Management Issues

There are many linkages among these management issues. For example, the destruction of critical habitats is often associated with overfishing. Pollution and coastal erosion contributes to the degradation of habitats. In Bolinao, the lack of political will is linked to the low level of environmental awareness. At both scales, overfishing is linked with the lack of alternative livelihoods. Thus, the management strategies being developed frequently address combinations of issues.

1. Overharvesting of reef resources and lack of alternative livelihoods

A casual conversation with resident fishermen reveals the sad state of diminishing catches. This has been validated by a fisheries resource assessment study (McManus *et al.*, 1992). If the reefs of Bolinao are to provide greater sustained production, new management measures must be adopted that reduce the fishing pressure. This requires the successful introduction of alternative livelihoods, both marine- and land-based.

Mariculture is now being field tested at an experimental scale with technologies developed or adapted by the Marine Science Institute. Land-based activities such as paper-making and soap manufacture have commenced in Bolinao. There is enthusiasm among coastal communities for these activities, although the programme is too young to gauge their long term impacts.

2. Lack of political will and the low level of environmental awareness

In developing countries, environmental issues are only now becoming important among politicians. In Bolinao, the degree of concern for the environment among ordinary citizens and local officials is variable. While one environmental problem may be perceived as important, another may be ignored even if the two are intrinsically of equal value. This is not to say that there are no informed individuals. Indeed, a number of residents, including some professionals, have demonstrated an appreciation for environmental issues. Unfortunately, they are in the minority. Yet full appreciation of environmental issues is central to any meaningful progress towards sustainable development.

A central objective of the CB-CRMP has been to increase understanding for environmental concerns, especially in relation to the marine environment. The project has launched a public information campaign and introduces members of the community to examples of successful resource management in other towns. These efforts have contributed to the election of a town mayor who campaigned on an environmental platform, and has opposed the proposed cement factory.

3. Degraded coral reefs

The desperation of those who must eke out a livelihood, or feed themselves from reef resources, leads to the use of destructive fishing methods which reduce the capital of the reefs to produce goods and services. Since fishing is essentially a hunting and gathering activity, it is of central importance to conserve the reefs and such associated habitats as seagrass beds and unvegetated soft bottoms. Yet it is extremely difficult, if not impossible, to preserve large tracts of the reef where subsistence communities are present. Thus, the approach thusfar has been to designate small areas as marine reserves or fish sanctuaries. The first concrete step was the designation of a small reserve immediately fronting the Bolinao Marine Laboratory. This occurred before the CRMP began and is believed to have had positive impacts on the communities near the laboratory. The Fisheries Stock Assessment Project (McManus *et al.*, 1992) recommended a larger marine reserve as a means of enhancing fish productivity but no action has been taken as yet. More recently, the CB-CRM project has introduced the concept of reserves to individual coastal communities. Three newly organized people's organizations have drafted proposals for small marine protected areas that have been endorsed by their village councils, and now await approval by the Bolinao Municipal Council.

3.3   Structures and Process for Scientific Inputs: Lingayen Gulf

The inputs of natural and social scientists to the Lingayen Gulf plan and the operations of the Gulf Commission have been significant. The generation and synthesis of information from 1986 to 1988 (Stage One of the ICM process) involved four institutions. The commercial and trawl fisheries was the research focus of the College of Fisheries of the University in the Philippines in the Visayas. Coral reef fisheries and resources and water quality were studied by the UP Marine Science Institute. The Bureau of Fisheries and Aquatic Resources of the Department of Agriculture evaluated the potentials and constraints of aquaculture in the gulf. The UP College of Social Work and Community Development conducted research to analyze the sociocultural, economic and institutional aspects of the environmental problems the gulf faces.

The information and ideas obtained through research were used in a planning exercise (Stage Two) that began in late 1988. Under the coordinatorship of the National Economic and Development Authority (NEDA) of Region I, task groups were formed to formulate sector-specific management programmes. Each task group drew together representatives of regional governmental agencies, local academic institutions and NGOs, and the researchers involved in the previous stage. The action plans were integrated into a comprehensive management plan in late 1989. In early 1990, the management plan was presented to representatives of the coastal municipalities and city surrounding the gulf in a series of town council meetings. The Plan was adopted and was incorporated into the Regional Physical Framework Plan for Region I for 1990-2020. However, until the Gulf Commission was created in 1994, there was no formal implementation of the plan.

The Commission has revised the plan to reflect a ten-year implementation schedule. Individuals with backgrounds in planning, environmental protection and management, resource rehabilitation and management, socio-economics and livelihood, serve as a Technical Secretariat to the Commission and representatives from academe serve on an Advisory Board.

3.4   Structures and Processes for Scientific Inputs: Bolinao

In Bolinao, scientific inputs are provided when the Local Government makes a request, or when studies indicate the need for management interventions. In the latter case, information is provided to local officials who do not always act on the scientific recommendations offered.

With the implementation of the CB-CRM Programme beginning in 1993, presentations on the need for coastal management have been made to Municipal Government and to the various village councils. Relationships, however, are informal and non-official. Unless an official working group is formed between the Municipality and the Project, scientific inputs will be channelled through public education, passage of municipal resolutions for specific management interventions, and as recommendations for land and coastal use planning. In 1994, when the proposal to build the cement plant was made, the institutions participating in the CB-CRM provided technical information that would allow the local constituents and their leaders to weigh the costs and benefits of the proposal.

3.5   Scientific Inputs for the Management of Lingayen Gulf and Bolinao

In this section, significant research findings which clearly defined critical management issues are discussed.

1. Capture fisheries

Silvestre *et al.* (1991) provide an assessment of the capture fisheries of the gulf. In particular, they conclude that the fisheries show biological overfishing with a high yield to biomass ratio of 5.2. For commercial trawling operations, the ratio is about 2.8. In both cases, the values indicate excessively high fishing pressure. There are an estimated 78 municipal fishermen per km of the 160 km coastline that operate 47 non-commercial boats and 26 commercial trawlers per kilometre. The use of 2 cm mesh size in the cod-end of bottom trawls leads to considerable growth overfishing such that 20 percent of yield and 40 percent of the value per recruit is lost. A mesh size of about 4 cm seems most appropriate given the level of fish production in the gulf (Ochavillo and Silvestre, 1991).

A study of the gillnet fishery of the gulf indicates that this is the dominant gear used by municipal fishermen in the gulf, and accounts for about half of the landings (Calud *et al.*, 1991). The non-enforcement of the 7 km from shore ban on trawlers has led to large overlaps in the fishing grounds of trawlers and gillneters. Catches by these two methods show heavy competition for the same species and size groups of fish.

To address this overfishing, Silvestre *et al.* (1991) have recommended the following: (a) imposition of a 7 km, 7 fathom ban on commercial vessels (current regulations limits them to 15 km from shore); (b) a 3 cm minimum mesh size; (c) no licensing of new trawlers and the prohibition of other gear such as Danish seines; (d) reduction of commercial trawlers by decommissioning old vessels; (e) reduction of municipal fishing effort.

To date, all measures except (e) have been adopted by the Commission but are yet to be enforced. The reduction of municipal fishing effort must await the passage of legislation by local governments.

2. Pollution and water quality.

An assessment of water quality in the gulf conducted by Maaliw *et al.* (1991) indicates parameters in the open gulf were within the limits set by the National Pollution Control Commission (NPCC). However, nutrients, heavy metals, suspended solids and coliform estimates were present at levels above the permissible limits in coastal and estuarine waters adjacent to the three major rivers that drain into the gulf. The Agno River showed the highest levels of lead and cadmium, while sediments from the Dagupan River were heavily contaminated with mercury. The Patalan and Dagupan Rivers showed elevated levels of coliforms, and both are classified as unfit for human use.

To date, there have been no actions to address water quality in critical estuarine areas, even though fishpond operations are a major source of revenues to Dagupan City. The source of heavy metal pollution appears to be a mining district in the Cordillera Mountains outside the jurisdiction of the two Gulf provinces. Pollution monitoring is included as a priority for environmental monitoring in the current ten-year master plan.

3. Economics and sociocultural dynamics of fisheries.

Añonuevo (1989) made an exhaustive study of the economics of soft-substrate municipal fisheries in the Gulf. He demonstrated that the net returns from fishing were low (an average of about Phil P 17.20 per day). Among the gear used, the most economically efficient are long lines and the illegal dynamite fishing. The least efficient is the bottom set gill net. He also noted that municipal fishermen average only about 11 days of fishing per month and are otherwise unemployed. Añonuevo has proposed that: (a) credit schemes to increase fishing capital and equipment should be stopped; (b) alternative sources of income should be developed to respond to the time not spent on fishing, as well as the availability of fishing family women and the children; and (c) to draw fishermen away from a dependence upon natural resources through appropriate forms of industrialization.

The initial attempts to implement the gulf-wide management plan has made the provision of new livelihood opportunities for fisherfolk a top priority. However, the manner in which these opportunities should be offered has not yet been agreed upon. Livelihood grants are being presented as economic opportunities that are not yet linked to mechanisms for reducing fishing pressure. The CB-CRM Programme is collaborating with the Commission to organize and link programmes to resource management.

Studies on the sociocultural dynamics of illegal fishing methods such as blast fishing and the use of poisons indicate the urgent need for public education (Galvez *et al.*, 1989). This is a slow process that promotes new values despite the relatively high returns of these methods and the practice of sharing such illegal catches among the community. The formation of local fisher organizations may be vital in creating social sanctions against such destructive fishing practices.

4. The status of coral reefs and associated fisheries in Bolinao.

A survey of the coral reefs of the gulf conducted by Meñez *et al.* (1991) showed that thirteen sites on the western side of the gulf had 30 to 51 percent living coral cover. Siltation from coastal activities and run-off as well as the use of dynamite and poison contribute to such degradation.

Evidence of overharvesting of coral fish include a decrease in adult fish density and in species diversity, as well as in the size of reproductively mature fishes. McManus *et al.* (1992) have documented such trends for adult fish communities along the slope of the Bolinao reefs from 1988 to 1991. For siganid fishes, the smallest recorded size of reproducing females was down to 3 cm, showing that intense fishing pressure has selected small and fast reproducing individuals.

These studies led to recommendations for the creation of protected areas, the development of alternative or supplemental livelihoods, and the promotion of public education to provide the needed social sanctions against economically efficient but illegal fishing methods. These recommendations have been included into the CB-CRM Programme and the Bolinao Municipal Government has acted upon some of them. For example, mariculture technologies for a number of species of the giant clam *Tridacna*, the sea urchin *Tripneustes gratilla*, and for some seaweed species like *Eucheuma*, are being utilized by local communities at a pilot scale. A portion of the animals are intended to be set aside for reseeding, and the rest will be used for food and to generate income.

4.   LESSONS LEARNED ON THE ROLE OF SCIENCE

4.1   Scientific Concepts and Information Useful to Management

The most useful inputs from both the natural and social sciences are those which define the management issues and how they may be addressed. For the Lingayen Gulf and the Bolinao reef system, critical scientific and cultural data that document the overfished condition of resources in the gulf underscore the need for an alternative livelihood base. It is crucial to recognize the poverty of fisherfolk in the formulation of any resource management plan and the imperative of developing alternative means of making a living.

Management recommendations were made by each scientific and ethnographic study that subsequently became the core of the Gulf management plan and the CB-CRMP framework. The recent actions of the Gulf Commission and the communities being assisted by the CB CRM project demonstrate that more effective governance is within reach.

In the specific case of the cement plant proposal, scientific information has been essential in informing the people of Bolinao and those living around the gulf of the need to protect their coastal environment. Thanks to the efforts of an informed and vigilant local group and its committed supporters, the proposal has not been approved. Nor has it been unequivocally rejected. The stalemate is evidence of a national development policy that has not recognized the full environmental impacts of such forms of economic development. Unfortunately, strategies to achieve economic development that will not further compromise the current state of coastal ecosystems in the Philippines have yet to be formulated.

Despite major scientific contributions on many topics, there remain important gaps in our understanding of the ecosystems involved. For example, the Gulf Plan would have benefitted from physical and chemical oceanographic data.

4.2   Role of Scientists and Institutions

While many believe that ICM is principally the domain of social scientists and government officials, it is often the natural scientists that initiate and sustain an ICM effort. This was the case for the Lingayen Gulf management initiative. Similarly, the CB-CRM is a joint undertaking of natural and social scientists.

The regionwide ASEAN coastal management project was a response to the realization that marine resources in Southeast Asia were under severe pressure and that scientists could contribute to addressing the root causes of these problems. The first task of natural scientists is to provide the objective data to support perceptions of resource depletion or degradation. Issue identification and analysis requires scientific data. Both programmes greatly benefitted from the pre-existing research conducted at the Bolinao Marine Laboratory.

During the research and planning stages of the Gulf Plan and the CB-CRMP, natural and social scientists played central roles in providing the motivation and ideas for ICM initiatives. These scientists had less influence on the formal adaptation stage and in mustering the necessary political action for meaningful implementation. In the case of the Gulf Plan, this resulted in several years of inaction. Once underway, however, implementation requires the contributions of scientists to further develop or adapt new management strategies and development technologies. Inputs from marine scientists are continually solicited by the managers.

In a developing country situation like the Philippines, the role of science in coastal management is a necessary but not sufficient condition for success. Recommendations for management must be solidly backed up by science because there are many political pitfalls to face. Pseudo-experts are sometimes used by opponents of sustainable development to resist the implementation of measures that are perceived as conflicting with personal interests. Hence, proponents of environmentally sound practices must feel confident of their arguments for reform.

Good science is not easy to come by in a developing country. It is often necessary to bring in outside funding to support local scientific work. Good science takes time to generate and ICM planning in developing countries may take longer than in the developed world. Yet there are probably no true shortcuts to this. The employment of expatriate manpower to generate local science is fraught with danger. Starting with the high costs of such expertise, the output is often of limited value because generalizations are made on the basis of short, and otherwise limited, observations. The acculturation process of expatriates is often not long enough for them to appreciate local conditions, although there are exceptions to this, as in the case of volunteer corps that spend sufficient time in their place of work to imbibe the local culture. The conclusions of short studies often takes a long time to correct. The better approach is to train an indigenous cadre of scientists who have firm commitments to see through local improvement.

5.   REFERENCES

Añonuevo, C., 1989. The economics of municipal fisheries: the case of Lingayen Gulf. ICLARM Conf.Proc., (17):141-55

Calud, A., E. Cinco and G. Silvestre, 1991. The gill net fishery of Lingayen Gulf. ICLARM Conf. Proc., (22):4-50

Galvez, R., T.G. Hingko, C. Bautista and M.T. Tungpalan, 1989. ICLARM Conf.Proc., (17):43-62

Maaliw, M.A.L.L., N.A. Bermas and R.M. Mercado, 1991. An assessment of water quality for Lingayen Gulf. Abstract. ICLARM Conf.Proc., (22):427

McManus, J.W., C.L. Nañola Jr., R.B. Reyes Jr. and K.N. Kesner, 1992. Resource ecology of the Bolinao coral reef system. ICLARM Stud.Rev., (22):117p

McManus, L.T. and T.-E. Chua (eds), 1990. The coastal environmental profile of Lingayen Gulf, Philippines. ICLARM Tech.Rep., (22):69p

Meñez, L.A.B., L.T. McManus, N.M. Metra, J.F. Jimenez, C.A. Rivera, J.M. Conception and C.Z. Luna, 1991. Survey of the coral reef resources of Western Lingayen Gulf, Philippines. ICLARM Conf.Proc., (22):77-82

National Economic Development Authority (NEDA), Region I, 1992. The Lingayen Gulf coastal area management plan. ICLARM Tech.Rep., (32):87p

Ochavillo, D. and G. Silvestre, 1991. Optimum mesh size for the trawl fisheries of Lingayen Gulf, Philippines. ICLARM Conf.Proc., (22):41-4

Silvestre, G., N. Armanda and E. Cinco, 1991. Assessment of the capture fisheries of Lingayen Gulf, Philippines. ICLARM Conf.Proc., (22):25-36

Reports and Studies GESAMP

The following reports and studies have been published so far. They are available from any of the organizations sponsoring GESAMP.

1. Report of the seventh session, London, 24-30 April 1975. (1975). Rep. Stud.GESAMP, (1):pag.var. Available also in French, Spanish and Russian

2. Review of harmful substances. (1976). Rep.Stud.GESAMP, (2):80 p.

3. Scientific criteria for the selection of sites for dumping of wastes into the sea. (1975). Rep.Stud. GESAMP, (3):21 p. Available also in French, Spanish and Russian

4. Report of the eighth session, Rome, 21-27 April 1976. (1976). Rep. Stud.GESAMP, (4):pag.var. Available also in French and Russian

5. Principles for developing coastal water quality criteria. (1976). Rep.Stud.GESAMP, (5):23 p.

6. Impact of oil on the marine environment. (1977). Rep.Stud.GESAMP, (6):250 p.

7. Scientific aspects of pollution arising from the exploration and exploitation of the sea-bed. (1977). Rep.Stud.GESAMP, (7):37 p.

8. Report of the ninth session, New York, 7-11 March 1977. (1977). Rep. Stud.GESAMP, (8):33 p. Available also in French and Russian

9. Report of the tenth session, Paris, 29 May - 2 June 1978. (1978). Rep. Stud.GESAMP, (9):pag.var. Available also in French, Spanish and Russian

10. Report of the eleventh session, Dubrovnik, 25-29 February 1980. (1980). Rep.Stud.GESAMP, (10):pag.var. Available also in French and Spanish

11. Marine Pollution implications of coastal area development. (1980). Rep. Stud.GESAMP, (11):114 p.

12. Monitoring biological variables related to marine pollution. (1980). Rep. Stud.GESAMP, (12):22 p. Available also in Russian

13. Interchange of pollutants between the atmosphere and the oceans. (1980). Rep.Stud.GESAMP, (13):55 p.

14. Report of the twelfth session, Geneva, 22-29 October 1981. (1981). Rep.Stud.GESAMP, (14):pag.var. Available also in French, Spanish and Russian

15. The review of the health of the oceans. (1982). Rep.Stud.GESAMP, (15):108 p.

16. Scientific criteria for the selection of waste disposal sites at sea. (1982). Rep.Stud.GESAMP, (16):60 p.

17. The evaluation of the hazards of harmful substances carried by ships. (1982). Rep.Stud.GESAMP, (17):pag.var.

18. Report of the thirteenth session, Geneva, 28 February - 4 March 1983. (1983). Rep.Stud.GESAMP, (18):50 p. Available also in French, Spanish and Russian

19. An oceanographic model for the dispersion of wastes disposed of in the deep sea. (1983). Rep.Stud. GESAMP, (19):182 p.

20. Marine pollution implications of ocean energy development. (1984). Rep.Stud.GESAMP, (20):44 p.

21. Report of the fourteenth session, Vienna, 26-30 March 1984. (1984). Rep.Stud.GESAMP, (21):42 p. Available also in French, Spanish and Russian

22. Review of potentially harmful substances. Cadmium, lead and tin. (1985). Rep.Stud.GESAMP, (22):114 p.

23. Interchange of pollutants between the atmosphere and the oceans (part II). (1985). Rep.Stud. GESAMP, (23):55 p.

24. Thermal discharges in the marine environment. (1984). Rep.Stud. GESAMP, (24):44 p.

25. Report of the fifteenth session, New York, 25-29 March 1985. (1985). Rep.Stud.GESAMP, (25):49 p. Available also in French, Spanish and Russian

26. Atmospheric transport of contaminants into the Mediterranean region. (1985). Rep.Stud.GESAMP, (26):53 p.

27. Report of the sixteenth session, London, 17-21 March 1986. (1986). Rep.Stud.GESAMP, (27):74 p. Available also in French, Spanish and Russian

28. Review of potentially harmful substances. Arsenic, mercury and selenium. (1986). Rep.Stud. GESAMP, (28):172 p.

29. Review of potentially harmful substances. Organosilicon compounds (silanes and siloxanes). (1986). Published as UNEP Reg.Seas Rep.Stud., (78):24 p.

30.   Environmental capacity. An approach to marine pollution prevention. (1986). Rep.Stud.GESAMP, (30):49 p.

31.   Report of the seventeenth session, Rome, 30 March - 3 April 1987. (1987). Rep.Stud.GESAMP, (31):36 p. Available also in French, Spanish and Russian

32.   Land-sea boundary flux of contaminants: contributions from rivers. (1987). Rep.Stud.GESAMP, (32):172 p.

33.   Report on the eighteenth session, Paris, 11-15 April 1988. (1988). Rep. Stud.GESAMP, (33):56 p. Available also in French, Spanish and Russian

34.   Review of potentially harmful substances. Nutrients. (1990). Rep.Stud. GESAMP, (34):40 p.

35.   The evaluation of the hazards of harmful substances carried by ships: Revision of GESAMP Reports and Studies No. 17. (1989). Rep.Stud. GESAMP, (35):pag.var.

36.   Pollutant modification of atmospheric and oceanic processes and climate: some aspects of the problem. (1989). Rep.Stud.GESAMP, (36):35 p.

37.   Report of the nineteenth session, Athens, 8-12 May 1989. (1989). Rep. Stud.GESAMP, (37):47 p. Available also in French, Spanish and Russian

38.   Atmospheric input of trace species to the world ocean. (1989). Rep.Stud. GESAMP, (38):111 p.

39.   The state of the marine environment. (1990). Rep.Stud.GESAMP, (39):111 p. Available also in Spanish as Inf.Estud.Progr.Mar.Reg.PNUMA, (115):87 p.

40.   Long-term consequences of low-level marine contamination: An analytical approach. (1989). Rep. Stud.GESAMP, (40):14 p.

41.   Report of the twentieth session, Geneva, 7-11 May 1990. (1990). Rep. Stud.GESAMP, (41):32 p. Available also in French, Spanish and Russian

42.   Review of potentially harmful substances. Choosing priority organochlorines for marine hazard assessment. (1990). Rep.Stud. GESAMP, (42):10 p.

43.   Coastal modelling. (1991). Rep.Stud.GESAMP, (43):187 p.

44.   Report of the twenty-first session, London, 18-22 February 1991. (1991). Rep.Stud.GESAMP, (44):53 p. Available also in French, Spanish and Russian

45. Global strategies for marine environmental protection. (1991). Rep.Stud. GESAMP, (45):34 p.

46. Review of potentially harmful substances. Carcinogens: their significance as marine pollutants. (1991). Rep.Stud.GESAMP, (46):56 p.

47. Reducing environmental impacts of coastal aquaculture. (1991). Rep. Stud.GESAMP, (47):35 p.

48. Global changes and the air-sea exchange of chemicals. (1991). Rep. Stud.GESAMP, (48):69 p.

49. Report of the twenty-second session, Vienna, 9-13 February 1992. (1992). Rep.Stud.GESAMP, (49):56 p. Available also in French, Spanish and Russian

50. Impact of oil, individual hydrocarbons and related chemicals on the marine environment, including used lubricant oils, oil spill control agents and chemicals used offshore. (1993). Rep.Stud.GESAMP, (50):178 p.

51. Report of the twenty-third session, London, 19-23 April 1993. (1993). Rep.Stud.GESAMP, (51):41 p. Available also in French, Spanish and Russian

52. Anthropogenic influences on sediment discharge to the coastal zone and environmental consequences. (1994). Rep.Stud.GESAMP, (52):67 p.

53. Report of the twenty-fourth session, New York, 21-25 March 1994. (1994). Rep.Stud.GESAMP, (53):56 p. Available also in French, Spanish and Russian

54. Guidelines for marine environmental assessment. (1994). Rep.Stud. GESAMP, (54):28 p.

55. Biological indicators and their use in the measurement of the condition of the marine eenvironment. (1995). Rep.Stud.GESAMP, (55):56 p. Available also in Russian

56. Report of the twenty-fifth session, Rome, 24-28 April 1995. (1995). Rep. Stud.GESAMP, (56):54 p. Available also in French, Spanish and Russian

57. Monitoring of ecological effects of coastal aquaculture wastes. (1996). Rep.Stud.GESAMP, (57):45 p.

58. The invasion of the ctenophore *Mnemiopsis leidyi* in the Black Sea. (in press). Rep.Stud.GESAMP, (58)

59. The sea-surface microlayer and its role in global change. (1995). Rep.Stud. GESAMP, (59):76 p.

60.   Report of the twenty-sixth session, Paris, 25-29 March 1996. (in press). <u>Rep.Stud.GESAMP</u>, (60). Available also in French, Spanish and Russian

61.   The contributions of science to integrated coastal management. (1996). <u>Rep.Stud.GESAMP</u>, (61):66 p.

62.   Marine biodiversity: patterns, threats and development of a strategy for conservation. (in press). <u>Rep.Stud.GESAMP</u>, (62)